# ELECTRICAL WIRING HANDBOOK

**Other TAB books by the author:**

No. 1245
$15.95

# ELECTRICAL WIRING HANDBOOK

## BY EDWARD L. SAFFORD, JR.

TAB BOOKS Inc.
BLUE RIDGE SUMMIT, PA. 17214

FIRST EDITION

FIRST PRINTING—AUGUST 1980

Copyright © 1980 by TAB BOOKS Inc.

Printed in the United States of America

**Library of Congress Cataloging in Publication Data**

Safford, Edward L
  Electrical wiring handbook.

  Includes index.
  1. Electric wiring, Interior—Amateurs' manuals.
2. Electric wiring—Amateurs' manuals. I. Title.
TK9901.S19      621.319′24      80-14360
ISBN 0-8306-9932-5
ISBN 0-8306-1245-9 (pbk.)

# Preface

Every book is an adventure. As we begin our exciting journey through this one, we want to examine, analyze and discover everything pertaining to electricity as used in your home, condominium, or apartment. Some ideas may even be appropriate for offices. There are some differences in the wiring requirements of commercial and residential buildings which will be discussed.

We will examine electrical wiring in detail. We want to know *why* it is done the way it is, as well as exactly *how* it is done by the professionals. Our sources of information have got to be the best in the world. They are the master electricians who have put in long hours of study and years of apprenticeship before they became master electricians on their own. They have to pass the most rigorous kind of professional examinations administered by the city in which they qualify for the *journeyman's*, or master electrician's license.

We also have to acknowledge another source of information for this work, that being the design engineers. These men draw plans, figure loads, calculate wire sizes and determine fixture, box and switch sizes, and types and panels to be used in fusing. These professionals have spent years in colleges and universities and are wizards upon graduation. Of course, the architects must be included among these professionals. They must know about lighting, decor, rules and regulations and many other things if we are to have adequate, safe electrical power and lighting in our homes, condos, offices and workshops.

Finally, we must add that there is much in the way of personal experience contained in the following pages. Like every home owner, condo owner or apartment dweller, we have had our share of electrical expansion desires, required repairs, replacement of fixtures, renovations, replacement of switches, wall plugs and sockets, motor replacement in air conditioners, and almost everything else concerned with electricity around the house and shop. We have had a whole world of problems with lighting and safety which we were able to solve.

If we were to give you a single concept which we have found to be most important in doing electrical work, we would say it is to be "methodical". There are rules which come from the regulations, and then there are "rules of thumb" which are used by all good electricians universally. There are good and important reasons for these rules and methods. We must know and follow them to the letter if we want to insure *safety*, correct operation of whatever is being operated electrically, and, finally, ultimate satisfaction with the knowledge of a job professionally well done.

First we will examine the basic types of circuits used, what electricity is composed of and how to handle it, some information on wires and how to get electricity to "flow" where it should and not flow where it shouldn't. We need to understand insulation—the types and when to use what type. Some ideas on tools to be used will be of value. As we progress on our journey through this book we will encounter more advanced concepts, ideas and situations. We hasten to add at this point, however, that if you are an experienced veteran in the use of electricity or electronics, you might want to briefly scan the earlier chapters and proceed to the more adventurous ones.

When we said it is necessary to be methodical, we mean in checking to see if power is off, making the right kind of wire connections every time, using the proper precautions on safety and rigorously observing them, using the right kind of joint connection insulating devices or material and doing it properly (not quickly) when sealing an electrical joint, and so on. Never take for granted anything about electricity. Be methodical, systematic and certain. It may save your life.

From time to time in the following pages, we will include a little bit of that mysterious element called *theory*. If we can get through it easily and painlessly, it will make our understanding of the *why* we do things as we do more perfect. Also, in some cases, we just cannot begin to understand how some devices work unless we can relate a bit of theory to their operation.

While reading this book, you might consult books which is appropriate to our subject; *Electrical Wiring and Lighting For Home and Office* (TAB book No. 671). It could be a useful supplement. Now let us begin this adventure.

E. L. Safford, Jr.

# Acknowledgments

I would like to thank the following people, organizations and manufacturers who supplied illustrations or provided information for this book: National Fire Protection Association, U.S. Department of Agriculture, Perry Homes, Radio Shack, Guy Fish, Bobby Love, Houston Lighting and Power Company, Houston Homebuilders, Don Harris Jr. and Richard Gauthier.

# Contents

# Dedication

To my sons, daughters, grandsons, granddaughters and to everyone who must do things of an electrical nature.

# Fundamentals of Electricity

Before you shoot a gun, you have to know which end to place against your shoulder. Before you do anything with electricity, you have to know what a *circuit* is.

## THE SERIES CIRCUIT

The most basic circuit used to convey electricity from one place to another is called the *series circuit*. It has a number of parts to it which are common to all home, condo and apartment wiring. These are shown in Fig. 1-1.

Let's spell out what we see in this diagram. First, there must be a *source* of electricity. This can be a battery or the lines feeding into the house from the light pole transformer, wherever it may be located. The *force* or *pressure* of this electricity is called its *voltage*.

It is the voltage or pressure rating of the *insulation* on the wires which governs whether that electrical current (flow) will stay in (on) the wires, or whether, like lightning, it will choose an atmospheric path of its own and become uncontrollable. Water pressure is good if it stays in the pipes. It can be dangerous and cause all kinds of damage if it escapes. If electricity escapes from its wires, it might cause much damage. Fire may result or someone could be electrocuted.

### Insulation

Thus, we find ourselves considering insulation. This is the protective cover around wires. We should realize that the higher the pressure (voltage), the better that insulation must be. Also, we might begin to anticipate that there may be different kinds of insulation depending on just where our wires are to be placed. If they are in a very hot location (oven area, etc.), the wires might require a special kind of insulation. If they are placed in a location where they might get damp or wet, that might necessitate a different kind of insulation which won't deteriorate under these conditions.

## Flow of Electricity

Next, we might consider the flow of electricity. Naturally you'd agree that if we want a stronger flow of water, we need a lot of current to make it work, then it stands to reason that we will need large wires to permit that flow. You will find many sizes of wires in a home, condo, shop, office or apartment. Each size has been specially selected to carry the current and withstand a given pressure in order to operate the "stuff" connected to its outlets. Any change in the "stuff" or the wire size will make it impossible for the electricity to do the job it was intended to do in that circuit. You cannot use any size wire in just any application when you use electricity.

Examine Fig. 1-1 again. See how the fuse or circuit breaker (which is a current or flow operated switch) is connected end to end (in series) with a manually operated light switch. Then the light element is connected so that one end is on the incoming wire and the other end is on the outgoing wire. Then the wire connects back to the source. In a series circuit everything is connected—end to end—and the flow must be through everything as shown.

## Circuit Breakers and Fuses

The *circuit breaker* is so important that we must spend some-moments thinking about it. It is a little device so designed that when the flow through it exceeds its designed limit, a magnet will energize. This, in turn, will cause some contacts to "open" and thus "break the circuit." The metallic parts of the circuit separate so the current cannot flow. But this happens automatically when an *overload* is present on that circuit. The circuit breaker is so designed, then, that you can go out to your circuit breaker box and *reset* the device by a slight physical motion. If the over *overload* is no longer present on that circuit, the device will reset. If the overload is still present, it will continue to *trip* "off" and you cannot

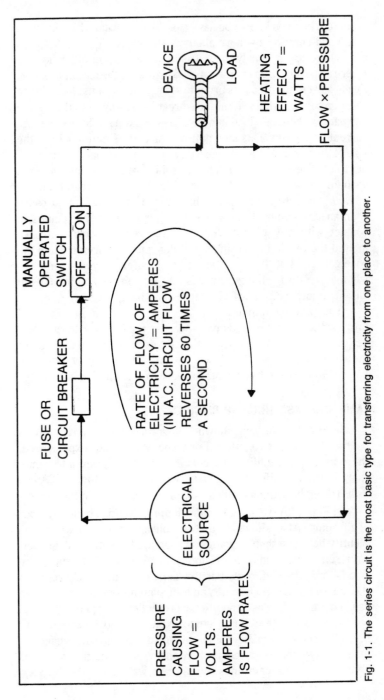

MANUALLY
OPERATED
SWITCH

OFF ☐—☐ ON

FUSE OR
CIRCUIT BREAKER

RATE OF FLOW OF
ELECTICITY = AMPERES
(IN A.C. CIRCUIT FLOW
REVERRES 60 TIMES
A SECOND

DEVICE

LOAD

HEATING
EFFECT = WATTS

FLOW × PRESSURE

ELECTRICAL
SOURCE

PRESSURE
CAUSING
FLOW =
VOLTS.
AMPERES
IS FLOW RATE.

Fig. 1-1. The series circuit is the most basic type for transferring electricity from one place to another.

reset it. Then you know you have a problem at hand which must be found and corrected—more about this later.

There are probably many cases, especially in older homes, where a fuse or group of fuses is used instead of the circuit breaker. Fuses and circuit breakers provide some safety against fires due to short circuits. A short circuit is a direct path away from the source and back without any load in between. A maximum flow of electric current will occur. This kind of flow is actually limited only by the size of the wire. When it is more than the wire can handle, the wire gets hot. Sometimes you can feel this warmth on the cord of a vacuum cleaner or an electric iron. If the flow is too great, then that *warmth* turns to actual *heat* which can be hot enough to cause spontaneous combustion of wood, paper or whatever comes in contact with the wire(s). Naturally the red-hot wire will have melted the insulation, which is not designed to survive when the wire gets red-hot. If a fire doesn't break out, and even sometimes when a fire starts, the wire burns into two pieces and breaks the circuit. Short circuits are dangerous and bad.

You may be thinking that the fuse or circuit breaker will prevent the short circuit damage because the fuse will burn in two before the wire gets hot. Also, the circuit breaker will trip before the wire can get hot. Notice we say hot. The wire may get warm before either of the two events happens. Your thinking is correct.

## LAMP AND PASS-THROUGH CIRCUITS

In Fig. 1-1 we have shown a diagram of how a series circuit operates and is put together. For some who are not experienced, there may be some difficulty converting that diagram into something in the real-life world. Thus, we proceed to Figs. 1-2A and 1-2B to see how things look as we might find them in the home.

In Fig. 1-2A we see a lamp with the switch as an integral part of the unit. Also shown is a cord consisting of two wires—no identity between them—going to a plug to fit into the wall socket. An electric iron, radio, television and hi-fi have this feature. The "loads" may be different, but the series circuit which operates them is the same. Sometimes on hi-fi systems and televisions we can find a small fuse as an additional protection if we know where to look. On most other home devices there is no fuse or circuit breaker. The system depends of that fuse panel circuit breaker or fuse to give the required protection to the circuitry, which is connected to the main lines through the wall plug.

If we have to repair units which are connected this way, it is a simple matter to unplug them and work on replacing parts. Cords often, through use, become brittle, frayed and dangerous. Plugs which become broke, bent and loose are also very dangerous.

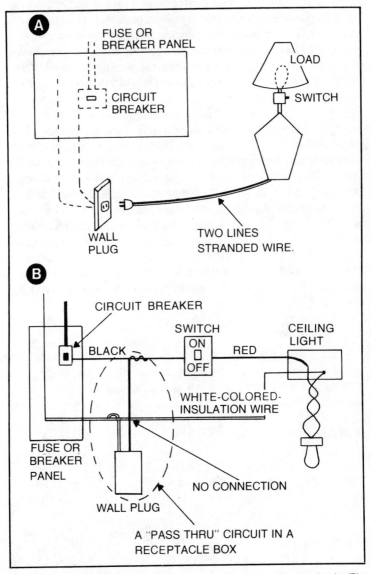

Fig. 1-2. Two circuits found in the home. (A) A lamp or device circuit. (B) Connecting a "pass-through" circuit.

17

There is no danger to us when we do make these repairs, providing we do them correctly. We are considering only the danger if the electricity is present. Of course, it is not dangerous if we have unplugged the unit.

In Fig. 1-2B we find the kind of series circuit which enables us to turn on a ceiling light, for example, from a wall switch. Notice the differences in wiring techniques. Here we have color-coded wires used to designate certain information about the circuit. We still have the same parts wired in the same way, although these parts are perhaps more durable, larger and of a different type than the one in Fig. 1-2A.

In Fig. 1-2A, using the lamp, we have stranded wires without color identity—just a pair of wires fastened together so they flex easily and bend without damage. They have a plastic insulation coating which may be clear, brown, white or almost any other color. They are satisfactory for the purpose intended for them, such as a light or whatever. A light or small load is intended as their end connection. Heavier cords with heavier insulation are designed for irons, large fans, and motors which also need flexibility in connection. Don't mix the two up. Never use a lamp cord to connect up a motor, electric iron or heavy duty hair drqer. The small wires can get hot, overheat, burn or cause damage. The wire size just isn't large enough to let the required amount of electricity flow, and you have trouble.

## SWITCH BOX WIRES

Now we come to a concept of having a wire that is "common" and a wire that is "hot." We refer again to Fig. 1-2B. Remember that the white wire is a common wire to all circuits. No switch, fuse or circuit breaker is connected to it in series or otherwise. It does carry electricity and can be very dangerous if you touch it and a water pipe or other "ground." The "hot" wire is the black wire coming from the fuse box or breaker panel to the switch or wall plug as shown. When the black wire goes to a switch, the other side of the switch is connected either to a black wire in the same kind of cable, or a separate black, red or blue wire to go on to the fixture. See Figs. 1-3A and 1-3B.

If you take the cover off a wall switch and look at the wires inside, you will probably see all the white wires connected together, or a white wire passing through the box, and colored wires connected together, or a white wire passing through the box, and colored wires connected to the switch. If two colors are used, then the black one will be on the side which comes from the circuit breaker panel on the house. The other color will be the "run" or line to the fixture.

Although we have shown just a wall plug connected to the line in Fig. 1-2A, Fig. 1-2B shows the line connected to a switch and a

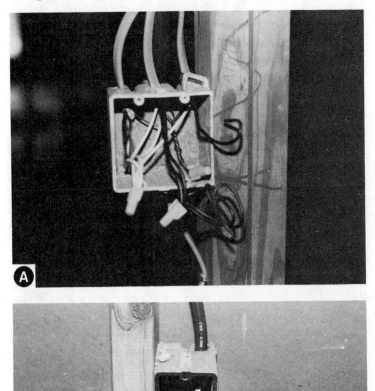

Fig. 1-3. Wires in a switch box. (A) Some wires are capped. (B) The center wire is the ground wire.

light. In many cases this extended line comes directly from a receptacle box. Here the common line can be connected to easily. Also, the hot side of the line (black) can be connected to the "pass-through" line through the use of the second screw on the receptacle. A pair of good connections thus can be made to the receptacle in the box for the extension. Now it is time to look at some wires associated with the concepts we have been discussing.

## Ground Wire

When you open the switch box, you find a wire which is either bare or colored green. That is the *ground wire*. Somewhere in the house it is connected to the water piping system. In most homes, it is also connected to a metal rod which is driven some 6 feet into the ground. Its sole purpose is to insure that all cases, pipes, etc., are at a common potential. It normally carries no electricity. If you touch it and a black or white wire, though, you might get a severe shock. Handled alone, it is not dangerous. When handled with any other wire, it can be dangerous.

The purpose of a ground is to prevent something you are holding from becoming charged electrically. For example, if you have an electric drill with a metal case, and somehow the wires carrying "juice" inside the drill become uninsulated and touch the frame, that is the same as through you grabbed a "hot wire" and were holding it. If you then touch a water pipe or are standing in water or on damp ground the electricity can flow from the frame and through you, into whatever you are touching or standing on. That might kill you.

When you use the ground type plug on your drill, and you have your wall plug properly connected to that bare or green wire, then a third circuit is completed which connects the frame of your drill, or whatever, directly to ground. Now, if your drill happens to short out connecting a hot wire to the frame, it will cause a fuse to blow or a circuit breaker to trip. But you will be safe. Never disconnect any grounding wires, and always connect them when they are specified on appliances or tools of any kind. That's what that bare or *green* insulated wire does in your home wiring system. Let's look at a plug which has a grounding connection in Fig. 1-4.

Unfortunately, not all wall sockets have that "third hole into which the grounding pin must go to make contact with the electrical grounding system of the home or office. Then the rounded pin on the plug of the machine is snipped off so that it can be inserted into such a wall socket. Of course, this eliminates the protection which was designed into the machine, and one does this at his own risk. Since it is a requirement that each box have a grounding wire inside it in some manner, we suggest an alternative to the snipping process. It might be better to buy a new wall socket which has the necessary three holes per plug. Then connect that grounding hole socket to the appropriate wire inside the box.

### Testing Series Circuit Wires

Every prudent and knowledgeable person who starts to do anything with an existing electrical circuit in a building or other place should test those wires to see if there is any electrical potential on them before starting to do anything. There are several ways of testing which are practical. A very simple one is to get a little neon bulb tester at most any radio store. Drugstores and

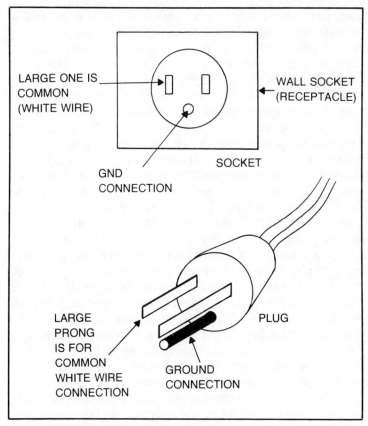

Fig. 1-4. A plug and socket with grounding connection.

sometimes grocery stores have them. This device has two leads and a neon bulb as diagrammed in Fig. 1-5.

Before trusting this device (or any device for that matter) to tell you if a circuit is hot or cold (hot is the case where electricity is present and cold is the state of no electricity present), you should test the device on circuits you know are hot and on one you know is cold. A wall socket can provide a hot connection. Just insert carefully the ends of the two leads into the socket holes. The bulb should light up. On a cold circuit you will have no indication of anything as the bulb won't light. This can be bad as you never know whether the bulb has burned out or whether there is actually no electricity present. So you need to test immediately before trying to determine the hot or cold status of a circuit.

You can use this kind of tester if you have nothing else to use. But it is not the best. Since neon bulbs require a given amount of electrical potential (volts) before they ignite, you might have a voltage present which is just below this level. Your bulb tester doesn't indicate this possibility. An electrical shock might be less. If you are sensitive to electrical shocks, it could hurt just as much as a higher level of voltage shock to a less sensitive person.

A second method of testing is to use just an ordinary lamp and remove the plug from the wire end. Or add a couple of wires to the plug, taping them so they won't come loose. We recommend removing the plug and twisting the stranded wires tightly so no loose ends are present. This will give you a kind of probe. Use it in the same manner as you would the neon bulb tester. Try it in a socket first. Sometimes you have to wiggle the probe wires a little to make contact; do so until you get the bulb to light. Then you know the lamp circuit is okay and its switch is "on." Be careful not to inadvertently turn the switch on the lamp off thereafter.

Now you can test the electrical potential on electrical wires in a box or elsewhere, but be advised that the same situation that prevails with the neon tester applies here. The lamp will not burn if the voltage is low. There can be a voltage present and the lamp won't tell you this. It will tell you if the expected 110 volts (or 220 using two lamps in series) is present. If somehow there is a reduced voltage present, which might come from a dimmer device on the circuit somewhere, then it will fail you. You are back again to the same situation as with the neon tester.

The best device to use is a small electric meter which can be purchased from a radio store. This will measure the voltage present from zero to whatever voltage you expect to have. It is quite

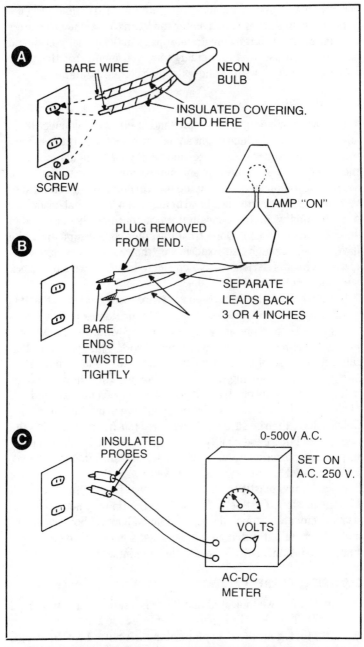

Fig. 1-5. Some voltage testing methods. (A) Neon tester. (B) A lamp turned to the on position. (C) An ac voltmeter.

accurate and easy to use, complete with full instructions. Even a person who has never read a meter can learn to use one in a matter of minutes. If you do any electrical repairs in the home, you should have one and use it. In an emergency you might use the other methods, but do so with an understanding of their limitations.

## ONE-HAND RULE

When you begin to make any kind of electrical connection to any circuit or wire which might somehow be hot, use the *one-hand rule*. Put one hand in your pocket and do all your manipulations with the other hand. It is difficult and almost impossible to get some actions accomplished. But it does prevent getting electricity through your body from one hand being on a wire, metal conduit or other ground while the opposite hand is touching the bare end of some hot wire (black). Don't trust the color code designation. We have found, through sad experience, that sometimes less-than-conscientious electricians have connected a white wire to a black wire for an extension of a "run."

With a bit of plastic covered two-wire (with ground so that it is really a three-wire cable) sheathed cable, or Romex as it is sometimes called, a person can easily connect the white and black wires properly and have an outlet which will "work." Just be aware that this can be possible, and don't trust the color code completely.

If you are removing wires from a box whether they are hot or cold, do it with only one hand. If you have to touch the exposed end of a wire, do it with one hand only. If you have to use both hands for the problem, then be sure one hand never touches anything but the *insulation*. Put a good quality rubber glove on one hand. If you are using pliers, a screwdriver or other tools, be sure that they have a well insulated handle, even if you have to wrap them with layers of black electrical plastic tape. Wrap a screwdriver right down to its tip if necessary. Do this so if you have the tip on a hot wire, you won't accidentally touch the shank against a metal box or conduit and a short circuit if that circuit happens, somehow, to get "hot." Figures 1-6A and 1-6B illustrate these points.

## ELECTRICAL CURRENT FLOW

In Fig. 1-1 we indicated that there is a flow of current around the circuit. The electrical energy must leave the source and return back to that source to complete the circuit. Two lines are necessary—one outgoing, the *hot* line, and one for the return current, which is usually called the *common* line. The hot line is normally color coded black; the return or common line is color coded white. The other important color is green, which connects everything metallic together. It is not used as a line to carry electrical current except in the case of a short circuit or some unintentional or accidental wiring mishap.

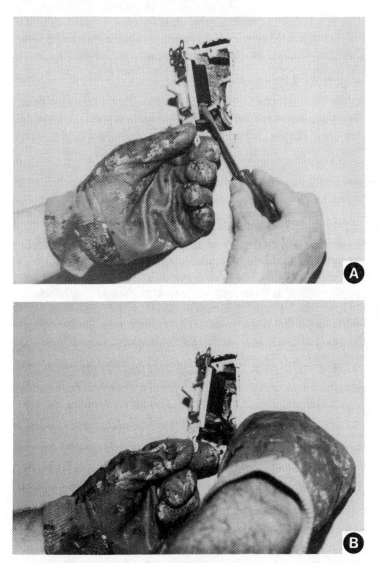

Fig. 1-6. The one-hand rule. (A) Unscrew the terminal. Put a rubber glove on your hand. (B) When using pliers to remove wire, use two gloves.

In Fig. 1-1 we show the current flowing in one direction but state that in an ac (*alternating current*) circuit, this flow actually reverses 60 times a second here in the United States. In some other countries the ac *frequency*, as it is called, may vary from this number, being as low as 50 Hertz (cycles) per second or even lower. We do not know of a higher frequency than 60 cycles being used at this time. This means we now introduce ourselves to the concept of a dc (direct current) and an ac kind of electricity flow. The dc flow is found in our automobiles. It comes from the battery and the dc generator. The ac flow is found in our homes and offices. So why the difference?

It has been discovered that it is *easier* and more *economical* (takes smaller wire size) to connect the electrical flow from the stations to homes, and around homes, than it would be to connect wires to permit the flow of dc in the same amounts and over the same distances. Also, it is easier to subdivide the voltage values of the electricity using alternating current than it would be to use dc for the same purposes. You can use a *transformer*, which is simply many iron laminations so shaped that wires can be wound around various parts of it to transfer the electrical energy in an ac circuit from one set of wires to another. You can make this transfer through a changing magnetic field, which occurs with ac but does not occur with a dc flow. Complete isolation of the wires is thus possible. Also, you can reduce the voltage pressure of the incoming line, which might be as high as 2,500 volts or higher, down to 110 volts, which is the value used in your home, or 220 volts to operate the stove or air conditioner. Most motors work with ac more economically than with dc.

As far as lighting is concerned, the dc will cause a light to burn just as readily as the ac does. You know that from your experiences with flashlights, batteries and your car lighting system. So, if we can reduce the bills and costs by using ac instead of dc for general purpose electricity, then it does seem important that we use it.

Now the ac can do some things that a dc cannot do. For example, ac can flow in circuits which dc cannot flow in because ac does its reversing flow. Let's examine a circuit which has an *open* in it as in Fig. 1-7.

There is a lot of surface area in the two sides of the *open* or break, as there might be if we took yards of tinfoil and placed the two layers a fraction of an inch apart. If we connected one side to one side of the line and the other side to the opposite wire, then we will have formed a *capacitance* across which the ac flows. But the dc

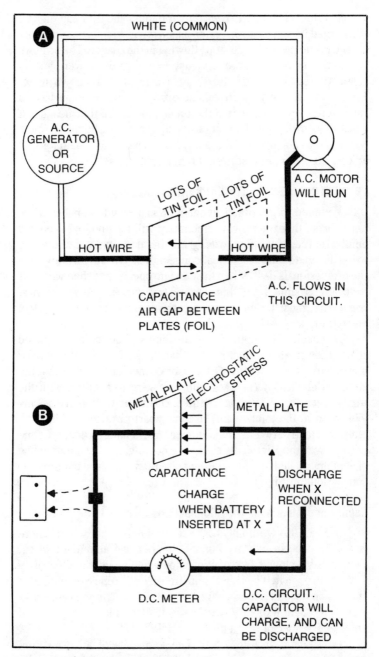

Fig. 1-7. (A) The motor will run as ac flows through the circuit. (B) In this example dc will charge the capacitor only.

does not flow. The dc will *charge up* the foils, and they can then be discharged in a reverse direction as shown in Fig. 1-7B. Once the charge is complete, the dc stops flowing in the circuit. The ac flows across the air between the foils just as though there were a solid connection between the foils. This can run a motor as shown in Fig. 1-7A. In fact, many electrical motors used in the home have a capacitor as a physical part of their makeup. Again, this means that when we consider the flow of electrical current, in a home, we must be careful to not accidentally provide a "capacitance" path for that ac current. It might just flow through that element.

## CONNECTING ELECTRICITY-CARRYING WIRES

We need to get practical again on just how to connect wires together so that the flow of electricity will be smooth, even and uninterrupted. If we do *anything* electrical around the homesite, we will have to make electrical connections. We should know how the professionals do it, when they do it properly, and then we won't have any trouble from this part of our system. Any electrical connection must be mechanically strong and must be as perfect electrically as possible.

Mechanical strength means the joint will not be easily pulled apart. Electrically perfect means that the surfaces of the two (or more) wires are so tightly wrapped or connected together that the atoms of electricity can pass from one wire to another, as if the wires were made into one solid metallic unit. If there is any looseness in the joint, it might come apart physically. Also, the atoms of electricity (electrons at least) might have a very hard time getting over the joint. This can cause heating, fire, malfunctioning of equipment, and other unpleasant situations. Let's examine some connections in Fig. 1-8.

### Copper and Aluminum Wire Connections

In Fig. 1-8A we simply connect two solid copper wires. There is a difference in how we connect copper and aluminum wires. Copper has a slower rate of expansion when heated by electricity flowing through it. Thus, an aluminum wire will tend to expand a lot and loosen its joint if it can. When it contracts due to the removal of electricity from its joint (switch off), it then is loose at that joint unless very special precautions are taken.

Look again at Fig. 1-8A. The two ends of an insulation-covered copper wire are stripped about 1 inch back from the tip. The wires are twisted tightly with pliers so that a good solid joint is made which won't become loose when it is moved or handled. If twisted too tight, the wires are stressed so much that the joint will break if any pressure or strain is put on it. Copper is somewhat flexible and will "give" to permit you to make a tight, twisted joint. Once this is done, the current practice is to cover the joint with a cap which has a metallic spring inside. It simply screws onto the joint and will provide insulation and help make the connection tight and firm. Such caps are obtainable in various sizes from electrical and hardware supply stores. Be sure to get the right size for your wires. See Fig. 1-8E.

This type of joint and the one shown in Fig. 1-8C are the most common ones used. In Fig. 1-8C the three wires are twisted in the same manner and a cap screwed over the exposed tips. If the joint is done correctly, there will be no bare wire visible at the large end of the cap. If there is, shorten the length of twisted connection and screw the cap on again, or get a longer and larger cap.

## Stranded To Solid Wire

You may have to connect a stranded wire to a solid wire when connecting light fixtures and various other devices to electrical circuit wiring. The stranded wires are found on the appliance or fixture. The solid wire is part of the wiring and extends out of an electrical connection box which might be plastic or metal. You have to twist the stranded wire tightly around the solid wire. Then use a good tight fitting connector cap to screw down on the joint to hold it in place and make it firm. It is very difficult and usually a bad practice to try to twist the solid wire around the flexible, stranded wire. The stranded wires will break and slide out of the "twist." Use a good cap or, if you want to have a very good solid joint, *solder* the two wires together.

## Tapping Into a Run

In Fig. 1-8B we show how you "tap" into a run or wiring. This is not common practice except when you are adding receptacles or switches to an existing wire run which is of the old style variety. That is, the wires are run in a parallel fashion on ceramic or equivalent type Knob and tube insulators in an attic or under a floor of an old home.

Always tape over these joints carefully using a good grade of plastic electrical tape. This type holds better than the cloth electrical tape, and you can wrap several thicknesses to get a really good amount of insulated covering over the joint. Be sure to use your

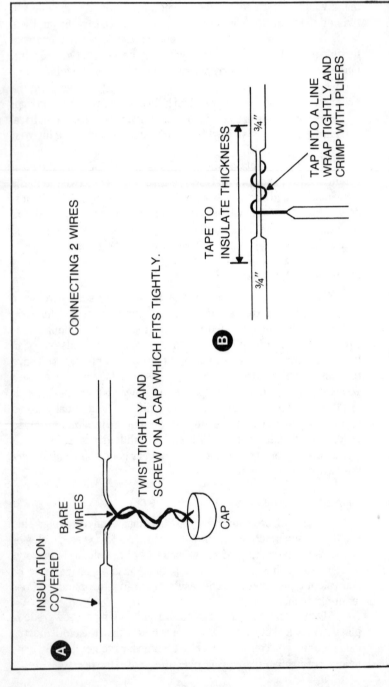

CONNECTING 2 WIRES

**A**

INSULATION COVERED

BARE WIRES

TWIST TIGHTLY AND SCREW ON A CAP WHICH FITS TIGHTLY.

CAP

**B**

TAPE TO INSULATE THICKNESS

3/4"

3/4"

TAP INTO A LINE WRAP TIGHTLY AND CRIMP WITH PLIERS

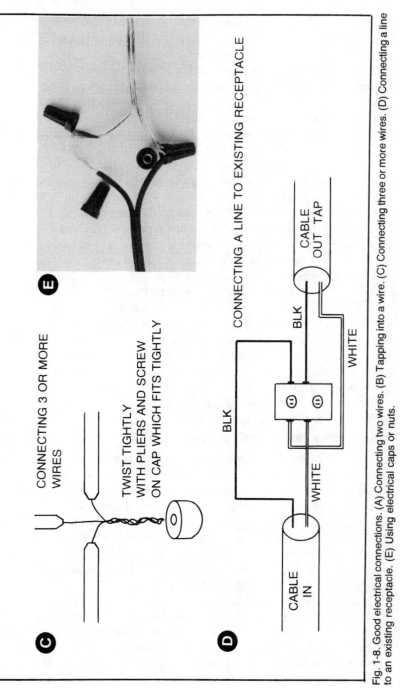

CONNECTING 3 OR MORE WIRES

TWIST TIGHTLY WITH PLIERS AND SCREW ON CAP WHICH FITS TIGHTLY

**C**

CONNECTING A LINE TO EXISTING RECEPTACLE

BLK

WHITE

BLK

CABLE OUT TAP

WHITE

CABLE IN

**D**

**E**

Fig. 1-8. Good electrical connections. (A) Connecting two wires. (B) Tapping into a wire. (C) Connecting three or more wires. (D) Connecting a line to an existing receptacle. (E) Using electrical caps or nuts.

pliers to make the tap-in so that the wiring is very tight. Pull the wires around the exposed connection point so the bare wires are as tightly bonded together as the added strength the pliers give will permit. Then squeeze the joint with the pliers to tighten it more if possible. Be careful not to damage the straight wire section so that it can break. The best way to make this joint is to wrap it as shown and then solder it with *rosin core* solder. If you are not experienced with soldering, then you can use just the mechanical connection if it is tight and firm. Check it to see if it is loose. If so, do it over.

A very common method of adding something into and out of an electrical box is to find a wall *receptacle* and go in there for the "tap-in." Here you have a ready-made screw connection which can be used. In Fig. 1-8D you see how one screw connects to the hot lead (black), and the same screw has another near it which usually is not used for any purpose. There is a small metal plate which connects the screws electrically and physically. If you place a wire under this second screw and tighten it down, you will have "tapped into" the hot lead at that switch or wall receptacle.

Be careful that you do not add a wire under a screw which already has a wire under it. The only way you can add a wire is to use washers between them. This is done so one wire doesn't pressure against another and flatten it out, spread it or cause it to become loose. The *gold* side of a wall receptacle or switch screw plate is for the hot lead always. The silver colored plate side of the wall receptacle is always for the outgoing or fixture wire. You also have to have a connection to the white wire in the box.

### Using Screws, Washers and Cap Connectors

You might use a machine screw, two flat washers, one spring compression type washer and a nut to connect two ends of aluminum wire. This will also work with copper wire. The wires are separated from each other by the washers, so one wire won't affect the other in expansion or tension. The screw is tightened until the compression lock washer is tight, which will give a permanent pressure to the connection. The washers should be at least ½ inch in diameter to dissipate heat as well as make a good connection. The joint should then be taped completely and thoroughly with many thicknesses of plastic electrical tape. We have used this kind of connection in a home with aluminum wiring when we replaced a regular wall switch with a *dimmer* type light control. In that case we had solid aluminum wire in the box to which we connected the stranded copper wire of the dimmer through this kind of joint.

We have seen some cases where light fixtures with copper stranded wires are connected to a solid aluminum wire using cap connectors such as we have illustrated and described. Since the cap connectors have a spring in them, they maintain sufficient tension to keep the connection tight. The method of such a connection is shown in Fig. 1-8E.

## USING LUGS AND CLAMPS

Some of you who are knowledgeable will be thinking about the use of lugs and clamps for electrical connections. We agree that these are good methods of making electrical connections if the wire is solid and of a large size. You will remember that when you get wires larger than size 8, they are usually made up of stranded elements. Otherwise, they would be too stiff to handle physically. You couldn't bend them easily around the turns which they must pass in ordinary wiring applications.

Let's look at some lug connectors and clamp connector ideas as shown in Fig. 1-9. There are many size lugs of the type shown in Fig. 1-9A. Simply get a size to fit the wire you are using. Remove the insulation from the tip of the wire back as far as that bare tip can be inserted into the sleeve. Using a crimping tool which is readily available from electrical stores, crimp the sleeve tightly down on the wire. In some cases you can insert stranded wire into the sleeve which will mash flat and tight against it, making a good electrical connection.

### Testing The Joint

You must always test the joint to see if it is mechanically secure. Pull on the lug with some force to see if it comes loose from the wire. If it does, do the job over. If you cannot get a tight enough fit to make the joint mechanically strong, then we suggest soldering the lug onto the wire. Just heat the sleeve with a soldering iron. Then run melted rosin core solder over the end of the sleeve onto the wire, and into the sleeve through any openings which might be there. You may not want to crimp the sleeve on the wire if you solder it. Rather, let the solder fill up the sleeve, making a strong mechanical joint as well as the best possible electrical joint.

The kinds of currents and the type of machinery or circuits the wire attaches to may govern just how you use lugs. If there is any possibility of a hot joint where the lug connects to a bolt or other screw terminal, then solder alone is not enough. If the lug gets hot enough, it could melt the solder. The wire would come loose or

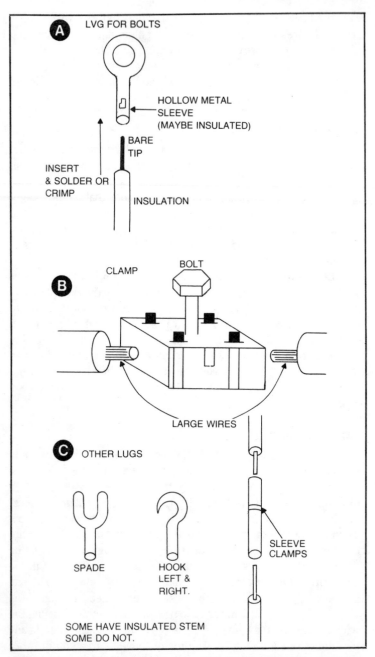

Fig. 1-9. Some lug and clamp type connectors. (A) You can use lugs as bolts. (B) Use clamps on large wires. (C) Other types of lug and clamp connectors.

34

make a bad connection and cause more heating. A loose wire might short out against other wires or the case or box, causing fire, blown fuses or the electrification of metal parts which should not have any electricity on them. Be careful. Make your joint mechanically strong. Solder it if you need added tightness, bonding or strength.

## Using Solder

Small lugs usually are not much of a problem unless you use them on stranded wires. Then you might need to solder the wire either before inserting it in the sleeve, or after, to make sure it is tight. You can solder the ends of stranded wires after twisting them tightly together and make up what will be a solid wire end to your wire. Insert this into the sleeve and crimp tightly. The solder will mash somewhat and make a good connection for you.

## CO/ALR Specification

If you think that you might use such lugs on the ends of aluminum wire (which is not recommended anymore for home wiring but is found occasionally in older homes), be very careful as

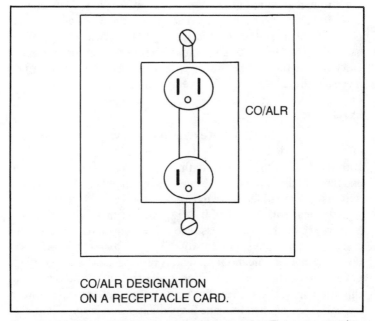

CO/ALR DESIGNATION
ON A RECEPTACLE CARD.

Fig. 1-10. The CO/ALR (copper to aluminum) designation. Fixtures or receptacles which connect to aluminum wiring should have this marking.

these lugs may be copper or of a metal which is not compatible with aluminum wire. Its expansion rate may be different. You could have a looseness or corrosion develop which would just mean more trouble and danger later. If you do want to use lugs on aluminum wire, or have to in some circumstances, then be sure you get lugs which are marked CO/ALR, which is the specification for connecting aluminum to copper or vice versa. If you use fixtures or receptacles which connect to aluminum wire, look for that marking as illustrated in Fig. 1-10.

Now consider the clamp as shown in Fig. 1-9B. This is found on the service connection to the house from the light company's pole, transformer or wherever large size wires must be connected together. The wires are not twisted. They are just run side by side (parallel) into the clamp. Then the screw or bolt is tightened down until a very strong and tight connection is made. Usually in the service connection you will find that these ends are taped tightly and thickly with a strong, well insulated electrical tape. Look for such joints outside on your home where the feeder wires come into the service pipe from the electric light pole.

These clamps can be found in various sizes and shapes. They are strong and do not use solder or any other type of bonding to make the electrical connection. The ends of the wires are scraped good and clean and then jointed and tightened down strongly. You have a tight and bright connection which is necessary for maximum efficiency in the transfer of electrical energy. It is also strong mechanically which is a safety requirement.

### EXTENSIONS

If you want to extend an extension cord, you will usually snip off the plug end. Then bare the two wires and twist them together with the ends of the added length. Tape up each wire connection and then put in a new plug. In some cases, simply take the old plug with its shortened leads, twist-connect the wires, tape them and you have made an extension. It works but, of course, this is not the best way to do the job. At least use a new plug on the ends so you don't have a double splice to worry about later, The job is very unsightly and very nonprofessional. A professional will first try to get a longer extension cord which is all one piece. If he cannot do this, he will get the parts, the plug, the socket ends and a long enough piece of wire to make a good extension cord.

If you are making an extension to some home wiring, you may be using the solid 14/2 or 12/2 designated wire. This means the

Fig. 1-11. A box is used for connection of extension wiring.

wire size 14 or size 12 and there are two such wires in the sheath, along with a ground wire (usually). Do not try to splice and tape as mentioned with extension cord. Provide a box connection into which the sheathed wires can run, and out of which the extension will run. Inside that box you will twist the ends of the black wires together, the ends of the white wires together and the ends of the ground wires together tightly. Tighten the connections using your pliers, and then put connecting and insulating caps on the ends of each twisted joint. Put all wires back into the box and screw the lid on. You may or may not use box connectors, depending on whether or not you are using a metal box (then you do use fittings) or a plastic box, where it is rather common not to use a fitting. If you are using a BX type wire covering, which is an armored metal stranded around the two wires, then you do need the box fitting to hold the BX in place.

Notice that you use a box to terminate the ends of the short run and the connection to the added extension wire no matter what kind of wires, conduit or cable you are using. Figure 1-11 illustrates such a box used for this purpose. This was in a garage where an extension was needed to make more electrical outlets possible.

## THE PARALLEL CIRCUIT

We have examined the series circuit and learned how we can put switches, circuit breakers and fuses in the black lead going to the load, which might be a light fixture, radio, television or wall receptacle. We learned that the white lead, which is the common one, is never broken to make provision for a switch or anything else. It is the return for the electrical current back to the source. In this kind of circuit everything connects end-to-end around a "loop" so to speak.

We now need to become aware of the *parallel* circuit which is commonly used with wall receptacles or room outlets, and which are usually all connected to common lines. Of course, there is a limit to how many wall receptacles may be connected to a common line. But there can also be several common lines which, in themselves, form parallel circuits. Let us look at Fig. 1-12.

In Fig. 1-12A we see how an incoming line with its white and black color wires connects to the sides of a receptacle. We have said that the white wire connects to the silver side of the receptacle and the black wire to the gold side. This is important. Another run of wire is also connected to these same colored metallic bars on the receptacle. Each receptacle usually has two screws on each side

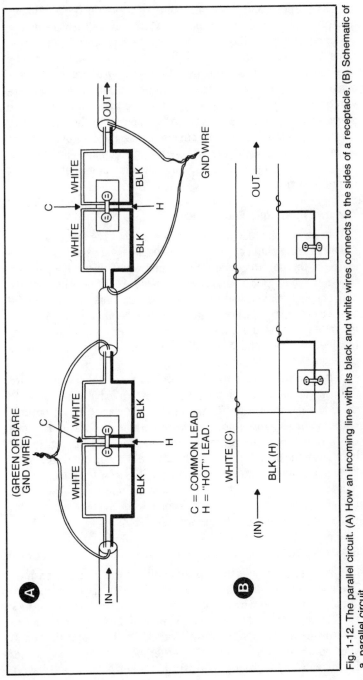

Fig. 1-12. The parallel circuit. (A) How an incoming line with its black and white wires connects to the sides of a receptacle. (B) Schematic of a parallel circuit.

39

for this purpose and the same color relationship is fastidiously observed. You never cross connect a white wire with a black wire.

The run of wire then goes on to a second receptacle where the same kind of connection is continued. The grounding wire may be bare or have a green-colored insulation. The wires are individually connected together by themselves, twisted with pliers and covered with an insulating and connecting cap as we have discussed previously. These wires never are connected to the receptacle, nor to any other part of the circuit unless the receptacle has a ground strapscrew and a third hole for connection to a grounding plug. Examine Fig. 1-13 to see how this kind of receptacle looks from the rear.

## Two and Three-Hole Receptacles

Look at your receptacles and see if there are three holes for the connecting plug or two. If only two, then you would expect the inside of the box to contain the ground wires twisted separately together and capped as described. If there are three holes for each plug connection, you will see, if you take the receptacle out of the box carefully, that a grounding wire has been connected to the grounding lug on the back of the receptacle itself. Instructions obtained with receptacles usually show whether or not a grounding wire should be so connected or not.

If it turns out that you have only a two-hole receptacle for ordinary lamps and you want to convert to a three-hole one to provide a ground return for something you want to run from this outlet, then get a three-hole or dual three-hole receptacle. Remove the power from the receptacle by opening the master circuit breaker or the switch at the fuse box. Carefully pull out the existing "two holer" and unscrew the wires from its screw terminals, one at a time. Connect each to the new receptacle. Be sure to observe the correct gold-silver bar code when going from one receptacle to the other. Wire them exactly the same.

## Grounding Wires

Now reach inside the box and pull out the grounding wires. You may have to add a short length of wire to the twisted group to get a small enough wire to fit under the ground screw of your new receptacle. If so, add this wire with caution. You don't want to destroy the connections of the other ground wires inside the box. If you find white wires connected together, do not tamper with these. Remember, your ground wire will be either bare or will have a

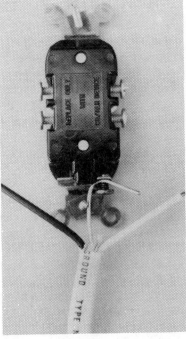

Fig. 1-13. A receptacle with connecting ground wire. The white and black wires connect to side terminals.

green-colored insulation. If you don't find such a wire in the box, then your home, apartment, or whatever just wasn't wired properly. You will not have a grounding wire to connect to.

If your home has metal conduit, you can ground to this with a short length of wire. Be careful not to let this wire touch a terminal screw on the receptacle. Otherwise, if you do not have a grounding wire as described, you are just out of luck at this location. You might examine other outlets and see if any have a ground wire inside. If not, then the only alternative is to run a single strand of about # 18 wire. You can use a lamp cord type dual wire with the ends twisted together to go from your device frame or ground strap to a water pipe. That will give you some electrical protection.

In Fig. 1-12B we show a schematic of how a parallel circuit looks. It is the same circuit as shown in Fig. 1-12A but just drawn differently. The drawing implies that a single run of wire, as shown at the "in"-"out" level, is tapped to make connection to the receptacles. That is not usually the case. The connections are made as shown in Fig. 1-12A. The screws on the side of the receptacles are indicated at C (common), which is a silver-colored bar, and at H, which is the gold bar for the hot black lead to connect to.

## CONNECTING HALF OF A WALL RECEPTACLE TO A SWITCH

In some cases a wall receptacle may have one of its two outputs connected to a switch. This is done so you can turn on a lamp, say, from a switch at the front door when you come into the house. Also, if you are adding some wiring to your workshop, you might want to have one outlet connected to a switch. Then you can connect a drill or something to this outlet and turn it off and on at another location, or with another switch than the one on the device. In some cases, the device may not have a switch. The only way to turn it off would be to unplug it. You may not always want to do that. Yet you do want to use a plug connection so that you can move or replace the unit if the case arises.

If you examine a receptacle carefully, you will see that most have a small metal strap which connects one screw area to the one next to it. This is important in parallel circuitry so that you can use the screws as shown in Fig. 1-12 to pass along the electric current. There is a complete circuit at H and C from the incoming line to the outgoing line.

If you take a small screwdriver or pliers and bend that connecting link between screws on the same side, then you will isolate each receptacle outlet, one from the other. This means you can run a hot line into one, add an extension to the line from that screw connection to run to a switch, and come back to the second outlet socket. Now you can have power to the second socket only when the switch is on. Of course, you won't change the strap on the silver side or common side of the outlet sockets. Leave them connected together. You will have a circuit as shown in Fig. 1-14.

Notice that you connect the incoming black wire to the screw at A. Also, a second wire must connect here, preferably a red one, but at least a black one as this is a hot line which goes to one side of the switch. The other side of the switch connects back to the socket to be controlled.

If it turns out that you need to go on from this receptacle to another one in a parallel circuit arrangement, you can only use the screw connection at A, and of course the white wire side of the receptacle. You probably should have lugs of the bolt connection type to make a good connection. In some cases, though, the three wires (black in, black out and red) might be twisted together around one wire, which is extended to fit by itself under the screw at A. This is shown in the insert of Fig. 1-14. Of course, you would tape up the wrapped wires to a thickness of two or three layers of electrical plastic tape. This is done so there can be no possibility of

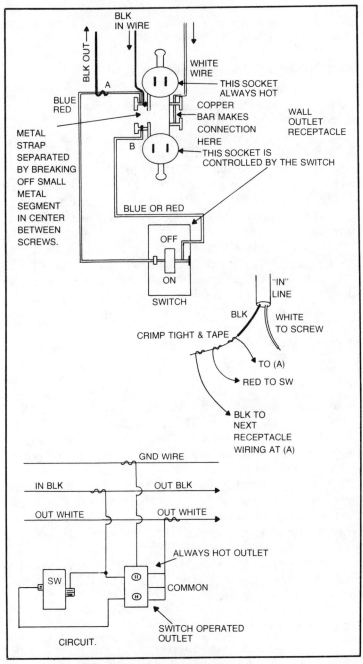

Fig. 1-14. Wiring an outlet receptacle to a switch.

43

having an exposed wire of this group touching any metal or any other wire in the box.

## Wiring Special Outlet Arrangements

You will find that there are special outlet arrangements which consist of the sockets, which are separate, and a small switch which fits on the plate right between them. With a little mechanical doing you can put together such an arrangement which will then give you a small switch and two receptacles all on the same outlet plate. These are reasonably strong when properly assembled and can give some flexibility to some kinds of wiring in the home, or modifications of the existing wiring. There are instructions with them so that you can put them together easily and wire them correctly.

The wiring of such arrangements may entail both the series and the parallel type of circuit. You now know that we do not necessarily have these circuits always unique and separate in a wiring system. You must be very careful that the circuit you are changing, adding to, or rearranging is of a type which you can identify. You would be in trouble if you tried to add some outlets in a series circuit, or tried to place a switch across a parallel circuit. Let us look at Fig. 1-15 and see what could happen if you make this mistake.

In Fig. 1-15A you see how the wires from an incoming pair are connected in series with a switch (the black wire) and then from the switch to one side of a lamp. The other side of the lamp is connected to the white common wire. This part of the circuit is okay. If the switch is turned on the lamp will burn, for the lamp and switch are in a proper series circuit.

Now examine the leg of wiring which runs from the switch to one side of the receptacle, and then from the other side of the receptacle on to the outgoing line. The grounding wire is properly connected to the grounding strap of the receptacle. Unless you have something plugged into the receptacle, there will be no *continuity* (electrical connection) between the line into the receptacle and the line going away from it. There will be no juice in the outgoing line under this condition.

## Dos and Don'ts

If you plug a device into that receptacle, then it won't work unless something is placed across the black and white lines going

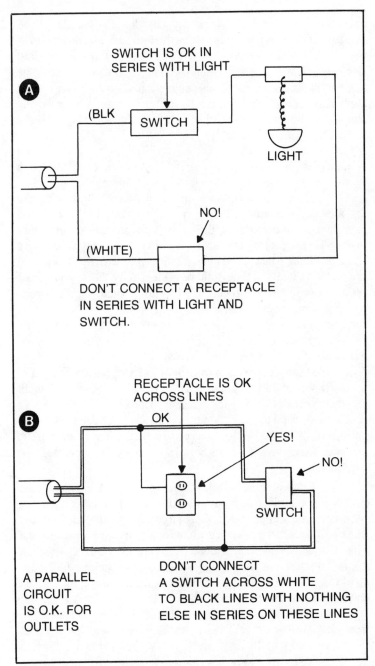

SWITCH IS OK IN
SERIES WITH LIGHT

**A**

(BLK      SWITCH      LIGHT

NO!

(WHITE)

DON'T CONNECT A RECEPTACLE
IN SERIES WITH LIGHT AND
SWITCH.

RECEPTACLE IS OK
ACROSS LINES

**B**    OK

YES!

NO!

SWITCH

A PARALLEL
CIRCUIT
IS O.K. FOR
OUTLETS

DON'T CONNECT
A SWITCH ACROSS WHITE
TO BLACK LINES WITH NOTHING
ELSE IN SERIES ON THESE LINES

Fig. 1-15. Errors in wiring. (A) Series circuit. (B) Parallel circuit.

out of the box. Then it will work only partially. The voltage will be divided between whatever is across the line later on and whatever is plugged into this receptacle. Let's say you plugged a lamp into this receptacle. If there was another lamp and switch across the line outside of the box past this connection, then each lamp would glow dimly in all probability. If you had an electronic device plugged into the receptacle, that device might be damaged by the low voltage-current available.

How might you make this mistake? You might make it if you thought that you could replace a switch *anywhere* in the house, shop or wherever with a receptacle simply by connecting the two switch wires to the two screw connections on the receptacle. Don't make that mistake. Remember, the receptacle *always* has a black wire on one side, a white wire on the other side and the ground wire (usually) in the center of the strap. A switch *always* has a black wire on each side, or a black, red blue or other color wire (except green) connected to it. Even though there is electricity there, you cannot replace a switch with a receptacle by simply changing the wires.

You can probably put a receptacle in a switch box by first removing the switch and connecting the two black wires together by twisting them together. Then make a connection from the white wire which is usually found in the back of the box to one side of the receptacle, and a connection from your twisted black wire joint to the other side of the receptacle. Of course, whatever outlet or light the switch controlled will be permanently on now, unless you have another switch on the device to turn it off. You can also remove the bulb (as from an overhead socket) or remove the fixture and tape up the ends of the wires inside its box. Tape them very completely and carefully.

Now examine Fig. 1-15B. Here is a very bad and dangerous circuit. If you replace a receptacle with a switch and then turn that switch on, you will provide a short circuit across the line from the white to the black wire. The result may be sparks, a blown fuse or a tripped circuit breaker. You might get shocked or hurt badly, and a fire might break out somewhere along the line back to the circuit breaker or fuse box. Never connect a switch from a black wire on one side (terminal) to a white wire on the other side (terminal). With this warning, and the explanation of the consequences, we trust that you won't get into trouble. Remember, the lines to receptacles are parallel lines, and the only way you can use a switch here is as we have already described. Use care and don't get hurt.

## VOLTAGE AND AMPERAGE

It will probably be informative to discuss the concepts of voltage, amperage and circuits in a home, office, condo, apartment, shop or wherever. The *voltage* is pressure to force the electricity to flow. We get that pressure in 110 volts for most home wiring or 120 volts if you want to be more specific. You will usually see that devices are rated 110-120 volts. This means the voltage can vary slightly, but not over 10 percent unless something unusual happens. You also know that for washers, ranges and some larger devices, the voltage required may be 220 volts plus or minus 5 percent or so. So the *pressure* on the electric wiring will be 110-115-120 volts, or it will be 220-230 volts, depending on what the wiring is connected to or put in the home to operate.

The *amperage*, however, is a different story. The reason we have lots of lines, fuses and circuit breakers in a home is because there is a limit to the amperage that a given wire can pass. Also, there are restrictions on how much current can be passed over a single circuit in a home or whatever. There are some very special rules according to the National Electrical Code on how many outlets can be on a given circuit. Remember that a circuit is one pair of wires coming in to the home from the fuse-circuit breaker box and going to a specified number of outlets or lights. Thus, you are going to have many circuits inside the walls and ceiling of your home, etc. You will have lots of wires (cables) running from the rooms to the circuit breaker box location. You will have parallel and series circuit connected to these runs. Figure 1-16 illustrates one situation as an example.

Normally a circuit in a home has only 15 amperes capacity. Some circuits for appliances will be at 30 amperes. Air conditioner circuits can be as high as 60 amperes capability to permit the motors to start. Motors may draw less current then than when they are running. The 220 volt pressure is used on air conditioner-heating circuits and large types of appliances. The other circuits in the home are normally on 110-120 volt levels.

When the circuit is for 18 amperes capacity, then the wire size is usually specified as 14/2, meaning a size 14 wire. There will be two of them, white and black, as well as the ground wire in the cable. If you have a BX cable or an aluminum or other metal conduit pipe, then just the two wires will be present for that circuit. It is entirely possible when using Greenfield armor conduit or pipe conduit to have many circuits which means many pairs of wires inside the metallic shield. In the non-metallic sheathed cable or lead-covered cable, just the two wires and ground are contained in the usual situation. But we have seen some cases where a multi-wire cable might be used in some living quarter applications, such as apartments.

If the circuit is for 30 amperes, then the wire size is usually specified as 12/2. For the 220 volt outlets and heavy duty appliance wall sockets you will find size 8/2. The 8/2 size has stranded wires to permit easy handling. The strands are pretty large, say about lead pencil thickness. Sometimes the grounding wire in this kind of cable is wrapped around the conductors. It forms a kind of metal sheath which is then covered with the plastic outside covering. See Fig. 1-16 and note the large white cable.

## PHASE

We would be remiss if we didn't discuss to some extent the *phase* of alternating current circuits. *Phase* is hard to understand if you aren't familiar with electrical theory, but it may not really be that difficult. When we think of phase in a practical manner we think first, perhaps, of motors. Motors which operate on alternating current are called single-phase motors, split-phase motors, two-phase motors, etc. A single phase motor, which is like an electric fan motor, is simply connected to a receptacle outlet or line with two wires just as a lamp is connected to those wires, outlets, etc. The wording single phase, if it is written on the label, really doesn't have much meaning for us.

### Split-Phase Motor

The same is true for a split phase motor. It also may just be connected to a receptacle or black and white wire line, and it will run when electricity is turned on. It may, however, have a slightly different physical appearance, such as a small cylindrical cannister mounted to its case. This is usually called a *capacitor* or *condenser*. It is the element which makes it possible to get motor armature rotation from the application of a single phase electrical current. The capacitor is the device which causes a split in this phase, and thus permits the motor to use the electric current to create the proper rotating magnetic field which pulls the armature around with it. There are some rare cases where this might be accomplished with an element called an *inductor*, but this is rare.

### Two-Phase Motor

Two-phase motors have three wires to connect to the three incoming lines (not two) upon which the electrical current travels.

Fig. 1-16. Multiple wire cables are needed for the required amperage.

What happens here is that there is a different time buildup of maximum voltage pressure on each of the two lines. The third is a common line. A rotating magnetic field is created in the motor windings and this, again, drags the armature around with it as it rotates. Two-phase motors are usually larger than the others we have mentioned. They are usually found in industrial applications or sometimes on large air conditioning units or large home heating units.

We need to point out that some so-called two-phase motors may actually run from a single phase ac input. They have one phase developed through the means of some reactive element such as an inductor or, more usually, a capacitor. This phase will "lead" or "lag" the reference phase of the incoming line. Such a situation is illustrated in Fig. 1-17.

## PRESSURE AND CURRENT BUILDUPS

Let us examine a graph which shows the buildup of pressure current in a single-phase, two-phase and a three-phase system (Fig. 1-18). At the top we see a single phase of voltage which, in this example starts at zero volts. It rises to 110 volts in a plus direction. Then it reverses and goes back through zero and on to 110 volts in a minus direction, and then back up to zero volts. That is one *cycle*. Since our voltage and currents are at 60 cycles per second, this must occur in 1/60 of a second. This is the way the pressure (volts) varies in out alternating because it goes from top plus to bottom minus over and over again.

We will assume in this example that the current flow will be in exact time with this voltage wave. We call this the "in-phase" condition. To imagine what the current is doing while the voltage varies in this manner, we simply think of the graph as representing also the current flow. Not that it also goes through zero and the two maximums (plus and minus), which here means the current reverses its direction of flow every 1/120 of a second or every half cycle.

### Two-Phase Representation

The two-phase representation as shown in Fig. 1-18 can represent the theoretical relationship of two *voltages* which are some 45 degrees *out-of-phase*. When we say out-of-phase, it simply means that the voltages do not build up and reverse polarity at the same time. We could say that the blue voltage is "leading" the black voltage, because the blue voltage gets to its peak first. Or we could

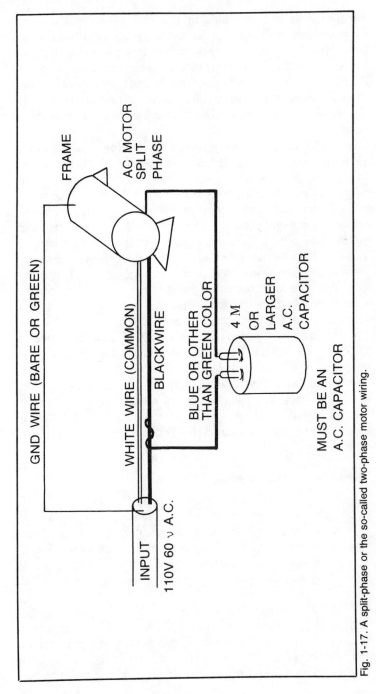

GND WIRE (BARE OR GREEN)

FRAME

AC MOTOR
SPLIT
PHASE

WHITE WIRE (COMMON)

BLACKWIRE

BLUE OR OTHER
THAN GREEN COLOR

4 M
OR
LARGER
A.C.
CAPACITOR

INPUT

110V 60 ν A.C.

MUST BE AN
A.C. CAPACITOR

Fig. 1-17. A split-phase or the so-called two-phase motor wiring.

say that the black voltage "lags" the blue voltage because it reaches its peak later than the blue voltage. We would be correct in describing the voltages in either manner. It only depends on which one we choose as the reference in order to say what the other one is doing with respect to this reference voltage. Again we assume that there is a current associated with each voltage, and each current is "in-phase" with its respective voltage. This is a *true* two-phase condition.

Now if we use a capacitor as shown in Fig. 1-17, then we would have the same kind of diagram but a different interpretation. In that case we would have the current *advanced* in phase with respect to its generating pressure (voltage). Thus, the blue curve of the figure would now represent *current*, and the black line the voltage in a single-phase system. You see how the current and voltage may not be the same in timing with alternating current systems. Under these conditions we have a split-phase arrangement rather than a two-phase arrangement of voltage. This is the kind of system most commonly found where the current and voltage of a single phase system are not in phase. Sometimes this is deliberate in order to run motors, and sometimes it is not intentional but due to the various kinds of loads which may be placed on the lines and outlets. Where the phases are not the same, the electric company will add elements like the capacitor to its lines. Sometimes you can see those little square or rectangular boxes on their light poles. The addition of these *reactive* elements, as they are called, will correct the phase difference and bring them more closely together in the time cycle. That is called "correcting the power factor."

### Three-Phase Arrangement

Also in Fig. 1-18 we see three voltage buildups in a three-phase system. We do find many three-phase arrangements used around industrial and manufacturing plants, etc., where large motors are used or high current requirements are found. Again, each phase will have a current associated with it. Each current may lead or lag its own respective voltage depending upon how that "phase" is connected to its loading device.

As we have seen, when we are dealing with a single-phase system which is what most homes, condos and apartments have as their electrical supply, then we use this directly for lights, stoves and water heaters. Or we can split its phase, causing its current to lead or lag the voltage pressure and get the kind of input needed to make some types of electric motors operate. If you have a heater or air conditioner motor which has a capacitor that splits the phase, this element changes with time so its "capacity" is less. Or it shorts out or otherwise fails. Then the motor to which it is connected just won't run. Sometimes just replacing that element for $10 to $12 may again make the motor run satisfactorily. You won't have to buy a $100 or $150 motor to replace the one installed.

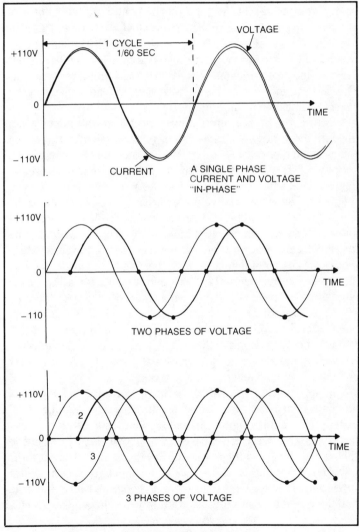

Fig. 1-18. Graphs of one, two and three voltage phases.

There is a way to test these capacitors if you suspect one to be bad. Any good radio shop can show you how to do it.

## POWER FACTOR

We will just mention a concept called *power factor* as this can be important sometimes in devices which use alternating current. The power factor describes the difference in heating value of an alternating current and a direct current of the same voltage pressure. If you pass direct current through a home stove element, that element has a resistance to the current flow and it generates heat because of this resistance. There is a very specific amount of heat which is related to the size of the resistance, the size of the current and the amount of voltage. There is no phase with direct current. The current and voltage are known to always be present in the same relationship all the time.

Now if you pass an alternating current through a motor winding with the voltage the same as for a dc you might not have the same amount of power. Why? The ac and voltage may not be "in-phase." The amount that they are not in-phase is measured in degrees, and a formula (EI cosine phase angle) will then give you the power factor, or relative power capacity, as compared to the direct current capability. E stands for voltage and I is current. We know that power can be rated in *watts*. Wattage is arrived at by multiplying amperage by voltage. If the phase angle is zero, which means the cosine of that angle is 1, then we have power factor as unity and the heating will be exactly the same for ac as for dc Notice, however, that if the phase angle is 45 degrees, then the cosine is 0.5 numerically. The heating effect of the ac is just one-half of that which would be produced if we applied the same voltage of direct current to the stove element. Stove elements are *resistances* and have unity power factors. Motors have *reactances* and have less than unity power factors.

Since we have brought up the subject of watts, let's spend a moment on this. You get a lamp at the store and it has a rating of 100 watts. You use this lamp on 110 volts and we assume a unity power factor. How much current will pass through that lamp? If you divide the wattage by the voltage you will find the current. Here it turns out to be .909 amperes. Remember this formula. It can be helpful at times to find the wattage, if you know the volts and amperes, or to find the current if you know the wattage and voltage. The formula is:

Watts (W) = (E) (I) cosine phase angle.

If you assume a unity power factor, you will be "in the ball park" by dropping the cosine phase angle term and just using E in volts and I in amperes.

## INPUT WIRES

When you consider the lines bringing the electricity into your home from the electric company, you will find that there are usually three lines which do the job. It makes no difference whether the

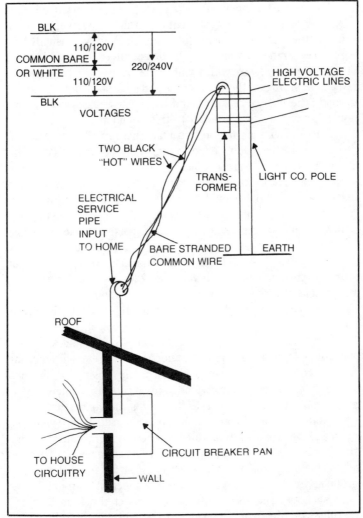

Fig. 1-19. Connecting the service for 110/120 and 220/240 volt capabilities.

lines go from the pole to a fixture on your roof, called the *service inlet*, or whether those lines come underground from a transformer to your circuit breaker panel on the side of your house. These lines feed through your meter into your circuit breaker box. The wires are large and could be from size 2 to 6, having a current capacity of form 140 to 80 amperes on each line. Notice that as the wire size number increases, its current capacity decreases. This is called an inverse relationship. The size 2 wire will handle more amperes. It is a larger size wire then the size 6.

Examining the input wires again, we find that two will be black normally. They can be any color but green and, of course, the color refers to the color of the insulation. The third wire will usually be a bare wire, stranded, which may be bright silver in color. It could also be a large size wire with a white color. Any wire which is white is a common wire. If the wire is bare and stranded, then it is also a ground wire or a common ground wire. All white wires in your home, condo, etc. will connect to this common wire.

The connections to the two black wires will be divided in order to make the loading even on the electric company's circuits. Basically you will want one-half of your house loading, or the current which is drawn when everything is on, to come through one black wire. The other half of the current should come through the other black wire. That is called *symmetry*. Also, we need to know that when the internal connections in the house are made from one black wire to the ground (white wire), the voltage pressure is 110-120 volts. This will be the same no matter which black input wire you are connected to. If you need 220-240 volts for anything which normally requires over 2,000 watts, then you will have a connection from one black wire to the other black wire at the circuit panel.

We can now generalize this situation by saying that we have a three-wire input at the *service* (connection to the electric company lines). From these three wires we can obtain either 110/120 volts or the 220-240 volts which we need in our homes, condos, apartments, shops, etc. Figure 1-19 illustrates the connections.

Notice that we do not use the common ground wire when we need 220-240 volts ac. Both lines will be black or a color showing that these wires are "hot" to ground. Great care must be exercised when obtaining 220-240 volts as this is a very dangerous level. The grounding wires are still used in the connections through three-hole sockets. The third hole is to connect the frames of the machines or whatever to ground, just as with the three-hole 110-120 volt sockets.

# Analyzing Some National Electrical Code Regulations

**2**

If you are thinking of becoming a master electrician and going into the wiring profession, then you must take an examination which covers to a large extent the National Electrical Code rules and regulations. You can find the complete code written in book form in many book stores, or they can order it for you.

If you are just an ordinary home owner or a person who likes to work with electricity, then you too should have a good knowledge of these rules and regulations as they pertain to the kinds of wiring in which you are interested. That may be for the home, condo, office, shop or wherever. The main reason for knowing about these rules and regulations is that almost everyone follows them, or at least they should. This means that if you are going to work in the electrical profession or do electrical work either as a hobby to save money, then you need to be able to converse with professionals. You must use some standard concepts as a basis for the discussion. The rules provide you with this common basis. This National Electrical Code is published by the National Fire Protection Association in accord with the recommendations of its technical committee. Safety and fire prevention are its primary goals.

Of course we cannot, by any stretch of the imagination, try to include all the rules and regulations. The diversity of electrical installations and individual situations, problems and requirements are almost beyond imagination. What we can do, however, is to look at and analyze some of those rules and regulations which fit into the scope of this book. Thus, we can give you a more complete picture of what is right and wrong than just by going on experience alone. We do need the benefit of experience and ideas from the professionals, but we will combine this with the basis for doing the job as contained in these rules and regulations.

It is informative to know that in most cities where there is a city electrical inspector, the rules and regulations are the basis for his approval or rejection, of a wiring installation. If you plan to do any such work, then a discussion with your own city electrical inspector might be worthwhile. He will tell you right away that a master electrician must do the work if he is to approve it. Even then you will know what kinds of things the Inspector would look for.

Some changes, additions and modifications to your existing wiring are generally within your own capability and, if done properly can be safe and give satisfactory performance. But if you want your home wired and insured, then it must be inspected by the city inspector and the work done by a licensed master electrician. The same is true when you completely rewire an older home. So use some judgment in this area as to what you can and should do and what should be contracted for.

## TERMS AND DEFINITIONS

We have said that a copper or aluminum wire will permit electrical current to flow through (over) it. We now recognize that there is another conductor called the copper-clad aluminum wire which is used in home wiring. This kind of wire has a copper covering bonded to an aluminum core so that many deficiencies of aluminum wire are overcome. These wire types are called *conductors*.

There are three general conductor descriptions which are stated in the National Electrical Code. The first is the *bare wire*. The second is the *covered wire*; note that the covering *does not insulate* or prevent the flow of electricity through the wire. And the third is the wire which has an electrical covering which *does insulate* the wire. Most commonly used are the bare wires, which we have said are the grounding wires, and the electrically insulated wires. You must be careful when using wires to see that the stated electrical pressure they will withstand (normally 600 volts) is much higher than the voltage pressure you intend to apply to the circuit using these wires. The 600 volt insulation is common in circuits carrying 110/120 volt pressures.

When using cap connectors sometimes called "electrical nuts" by the professionals, you must be certain that the metal inside is the same kind of metal as the conductors. If it is not the same, there might be oxidation or some chemical action between the different metals which would cause a bad electrical joint and result in *heating*. This could lead to problems such as loss of electricity in the circuits and possible fire.

We will often use electrical wiring in a *damp location*. This is defined by the code as being under canopies, roofed open porches and places where dampness or wetness can be present on the wiring. Some inside locations such as basements and cold storage

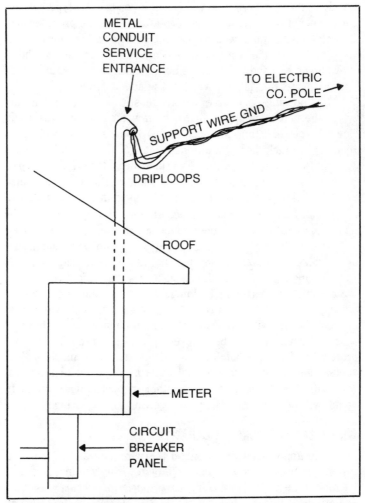

Fig. 2-1. Drip loops on the service.

areas like sheds, barns and tool buildings also will permit the accumulation of moisture on the wiring. Special damp resistant conductors and cables need to be used in these locations. Just ask your electrical supply house for a damp resistant cable.

We have seen homes wired where a *dry type* three conductor (two wires plus ground) is extended outside the front or back porch roof to provide a light or convenience outlet. This is bad. You should always get that special damp or waterproof type conductor cable for that particular run outside of the house proper. Sometimes a line is run from the house to the garage. It is overhead and you don't think any problem will result if you use dry type wiring. That may not be the case. Unless good *drip loops* are provided and the cable insulation is waterproof, you could be asking for trouble (Fig. 2-1). Use the waterproof and moistureproof cables where they should be used. The cost is not so much that you can't do it, and the protection is well worth those added pennies!

It is possible, of course, to have plastic or metal conduit into which the wires can be run that is both water tight and gasproof. Special types of fittings are necessary over the switches and lights in areas in which these conditions exist. You may not be involved with this kind of situation unless you provide underwater lighting in your swimming pool or something similar. But be aware that conduit exists which can and should be used in these situations.

As long as we have stated, according to the code, the definition of *damp* locations, we need to include the definition of *dry* and *wet* locations. A dry location is normally is not subject to dampness or wetness, however it *may* be subject to some dampness or wetness under special conditions such as when you are constructing a home or apartment building. A *wet* location is a location underground (in earth) or in concrete slabs and masonry which has a direct contact with the ground and is in a place where there is water. The wiring might be exposed to weather directly and has no protection if run in that location. Protection here means being contained in a conduit of a special waterproof type, a sealed raceway (box type conduit) or some other kind of outer protection which will prevent the water from getting to the wiring.

## WIRE SIZES AND CURRENT CAPACITY

It is important to understand the wire size and its current capacity in order to have proper safety in wiring you install in any location. The current capacity, *ampcapacity* or *ampacity* as it is sometimes called, is shown in Table 2-1. Of course, the capacity

Table 2-1. Amperage Capacities of Electirc Wires.

| Wire Size (AWG) | Copper | Aluminum | Copper clad Aluminum |
|---|---|---|---|
| 18 | 10-21 | not given | not given |
| 16 | 15-20 | not given | not given |
| 14 | 25 | not given | not given |
| 12 | 30 | 15-25 | 15-25 |
| 10 | 30-40 | 25-30 | 25-30 |
| 8 | 40-50 | 30-40 | 30-40 |
| 6 | 55-70 | 40-55 | 40-55 |
| 4 | 70-90 | 55-70 | 55-70 |
| 3 | 80-105 | 65-80 | 65-80 |
| 2 | 95-120 | 75-95 | 75-95 |
| 1 | 110-140 | 85-110 | 85-110 |
| 0 | 125-155 | 100-125 | 100-125 |
| 00 | 145-185 | 115-145 | 115-145 |
| 000 | 165-210 | 130-165 | 130-165 |
| 0000 | 195-235 | 155-185 | 155-185 |

can vary with temperature, but we do not consider the temperature variations in a home, to be so great as to make a marked difference in conductivity.

While we will show the capacity for aluminum wire, we *do not* recommend this kind of wire for any home, condo, apartment, office or shop. *We recommend copper wire*, but not the copper-clad wire. Since many homes have been wired with either aluminum or copper clad wires, we need to include their capacities in the table for completeness.

We have shown a range of amperes for each wire size according to its metallic composition. This range is governed by the type of wire used, the lower range using types RUW, T, TW and UF, and the higher range using types TA, TBS, SA, ABV, SIS, FEP, FEPB, RHH and a few others. These examples are enough for you to specify a type of wire to your electrical supply house which, along with the current capacity you desire, will enable you to do the kind of work you need to do. These specifications are from the National Electrical Code.

## GROUND FAULT INTERRUPTER

It is now a requirement that a ground fault interrupter be installed in locations where there may be water pipes. This device disconnects the electrical circuit to the load when the leakage or fault current to ground exceeds some predetermined value which is less than that required to cause a circuit breaker to operate. In essence, this device is sensitive to small current leakages and will cause the circuit to open when such a leakage exists. It can be reset by pushing a small red button on its case. We found many of these installed in bathroom areas where there might be a change of getting the water pipes electrified, or of getting some device or someone connected between an outlet in that area and the water pipe. That could happen if someone used a hair dryer or an electric shaver which somehow developed a short connection between its wiring and its case; or if a cord became old and frayed and someone was touching that cord while at the same time touching the metal water pipes. Ground fault interrupters are essential units to have for safety. Many city codes require them.

The code specifies that all wiring and electrical work shall be installed in a neat and workmanlike manner. You can often determine when a amateur is doing the work by the sloppy way in which connections are made and wires are run. A real professional takes a great deal of pride in his work. When you view the results of his efforts, you find evidence of careful work which gives you confidence that the work has been done right. We interviewed a master electrician on the job who was doing a complex bit of wiring in a home under construction. We pulled out the mess of wires tucked into a switch box to ask a question about them and were pleased to see how quickly the man carefully separated and coiled the wires and put them back in place in a neat manner, even as we were talking. It was a habit, through long practice, for him to be neat and exact in his work.

## MOUNTING OF EQUIPMENT AND AIR SPACE

The mounting of equipment is directly related to the workmanship of the personnel on the job. Wooden plugs driven into masonry, plaster or concrete will shrink. When they get wet, they will rot. It is very important to use *approved methods of mounting* all kind of electrical fixtures, boxes, and appliances. Some special anchoring devices may be necessary. "Molly" and "toggle" bolts may be very useful in plasterboard to give a good solid installation. Most electrical stores have devices to mount boxes, conduit and fixtures.

It is always necessary to provide air space around any electrical device which is installed so that air can easily circulate around and cool the device. Never mount electrical equipment so close to wall, ceilings or floors that it cannot get this circulation. Remember, electricity does generate heat. The larger the devices, the more heat is generally produced. In some large installations it may be necessary to provide extra cooling to keep the temperature to a safe level.

## ELECTRICAL CONNECTIONS

Because there may be a chemical action taking place, the code says you must *never* connect two wires with dissimilar metallic compositions together (such as aluminum and copper or copper-clad aluminum) in a splice or on a terminal, unless the method of connection is approved for that purpose. It is always essential that the proper type connectors be used for the type (composition) of wire being used.

Be sure that connections to terminals do not damage the wires. When putting a wire under a screw, always place the wire so that the pressure of the screw will tend to wrap the wire tighter and not loosen it. It should be coiled on the screw in the same direction the screw or bolt is turned to tighten it.

When you have to connect two or more wires together, it is important to know that there are electrical connectors which are designed to make a very tight connection. These are called clamp or pressure connectors. When you know about them and examine them, you will be able to find the kind and type suitable for your use.

### Splicing Wires and Soldering Connections

The code says that you can splice wires together if you do it in the approved manner and in the proper places. Proper places means that the splice you make should be done in a box. If it has to be made somewhere in the "run" (length of wire or cable), then you need to be sure it is accessible and that proper precautions are taken to make the joint mechanically firm and tight. Normally you do not want a spliced connection resting against a wood beam, floor or anything that might catch fire. If that joint somehow gets loose or corroded, it will get hot due to the electrical resistance which then develops. Figure 2-2 shows one good splicing method.

The sheathed cable into which the tap or splice is to be made should loop out from its wood or wall support so that the loop is in air. Then the white and black wires are made bare at least 1 inch from the ends. The ground wire is usually bare to begin with. Then the sheathed (cable) conductors to be connected have their ends scraped clean of insulation. The white wires are firmly wrapped around the white wire of the "in" cable wire. The black wires are wound around the bare end on the black "in" wire, and the ground white wire connections must be thoroughly taped with at least three thicknesses of good electrical plastic tape or capped with electrical "nuts." Insulated staples can be positioned so that all three nonmetallic sheathed cables are firmly held and no strain will be placed on any joint.

To really do a good job, you should solder the connections using a resin core solder. If you do not know how to solder, you can use the pressure-twisted wire connection if you twist good and tight using your pliers. The code says the plastic tape wrappings must be as thick as the original insulation and as good an electrical grade as the insulation. Most plastic electrical tapes are rated at 10,000 volts insulation. They probably exceed the insulation of the wire itself. Use of "caps" simplifies the insulating of the connections.

### Conductors and Grounds

It is important that you use the proper connectors if you use the relatively new copper-clad aluminum wire. This kind of wire has been approved under the National Electrical Code for use as an electrical conductor. Ask the people in the electrical shops about the connectors for this type of conductor.

The code states that all premise wiring shall have a grounded conductor. This conductor shall be white or natural gray. We have considered these colors to be for the common wire in an installation. The ground wire is bare, having a green insulation. Now how does all this fit together when it seems there is a contradiction? Actually there is no contradiction. The common wire (white) which we have discussed is actually connected to a ground or common terminal somewhere in the electrical system. The bare wire may be connected to the water piping system in the home and also to a ground stake which is driven some 6 feet into the earth. This ground stake is metal and a good conductor. Normally, if you connect a meter between the white wire and this ground wire, you will not find a voltage present there. Sometimes you may find some voltage, but it will probably not be on the order of the supply voltage at 110-120 volts or 220-240 volts.

Do not be dismayed that the code states a grounding wire should be identified by a white or natural gray color. Remember that while this is slightly different in concept from our common wire designation and in accord with universal electrical wiring practice, the white wire in an electrical wiring system *is* grounded *somewhere*, usually at the electric company's grounding system.

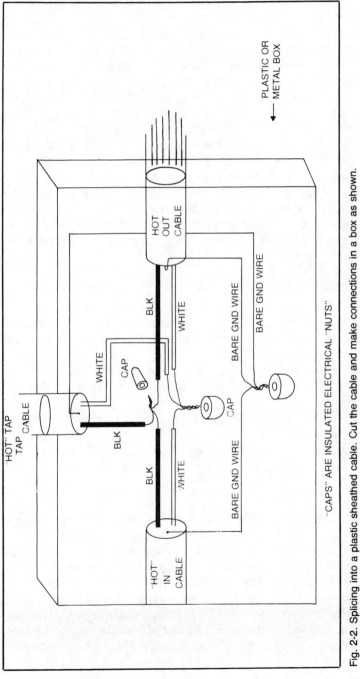

"CAPS" ARE INSULATED ELECTRICAL "NUTS"

Fig. 2-2. Splicing into a plastic sheathed cable. Cut the cable and make connections in a box as shown.

The bare wire is earth-connected and water pipe-connected right there at the premise location.

We need, here, to insert a note of caution. The code is so written that under some circumstances and under certain circuit restrictions the black and white wires may be interconnected. One example is when two or three wires are used for a single-pole or three-way switch. In this case you should always test the white wire to ground to see if it is connected to a hot wire somewhere prior to the point at which you are entering the circuit. This is sometimes done perhaps inadvertently, accidentally or from ignorance. If you are just assuming that always the white wire will be the common or grounding wire and that it has no potential to a conduit, water pipe or earth itself, you could be wrong and get severely shocked. Always turn off the main power switch (breaker) to the house, etc., before going into the wiring. Even then, test with a meter or suitable tester to see that the circuit you are going to work on is dead.

## Avoiding Mistakes

We have personally used a lamp (neon) type tester as well as a voltmeter (150 volts ac type) to measure from the white wire side of a receptacle to the conduit or ground. This is done to see if there was a mistake in wiring, like connecting a white wire to a black wire somewhere. Testing with caution when the circuits are hot can be helpful before making splices or adding fixutres to an existing system.

The code says only in the presence of a qualified person can terminals for grounding conductors be permanently identified at the time of installation by a white marking. This is to prevent an error in marking of terminals by inexperienced personnel.

## Sockets and Leads

It is important that a light socket of the common type found on lamps has its *shell* connected to the common or grounded wire of the electrical system. The small connection to the base of the bulb in the center of the socket is to be connected to the hot lead. This will be especially true regarding the installation of light fixtures in a home, condo or office. It cannot always be true with a lamp in the home because a non-identified two-wire cable is used. The plug may be inserted into the receptacle in either of two directions. One way will make the shell of the socket hot. The other way will make it cold and connected to the common lead as it should be.

If you use a test socket which has leads coming from its plastic compositions you will normally find that one lead is white. That is the lead to the shell of the socket. It should be connected to the white lead of the home wiring system. Remember this if you use such a lamp socket for testing for voltages.

Appliances leads must be so identified that you can always tell which lead is the grounding or common lead (white lead). On receptacles you can find those which have a definite polarized socket, one side larger than the other, which will identify which side of that receptacle is connected to the white wire (common wire) of the electrical system.

## INDIVIDUAL INTERPRETATION OF THE CODE

It probably seems that we have obscured the real method of wiring with some code statements which are confusing. We do not intend that this be so. Actually there is no official interpretation of the code. It consists of many guidelines which then must be interpeted by the electrical inspector of the city who inspects and approves electrical wiring jobs. This makes it difficult for the home owner or individual who wants to add, modify or change his home wiring. He looks for some concrete examples of what to do and finds, perhaps, some confusion, especially if he gets a copy of the code and tries to interpret it himself. Accepted practice is the basis for good wiring, as well as code specifications.

We have stated that the white wire in electrical sytems is the common conductor. That means it is the essential part of the electrical circuit which provides the return to the source for electrical currents. It is usually connected to earth somewhere outside the home. This is done by the electric company. In the fuse or breaker box, this wire connection will be found connected to a bare wire which comes from the electric company lines. Throughout a home, condo, apartment, office, shop or wherever, it is systematic, safe and logical to connect all white wires together and not break them with switches, receptacles or anything else.

The code has referred to the grounding wire on appliances as being the white wire. As we have shown, it may be true that somewhere the wire is connected to ground. In normal installations there will be a separate bare or green wire on the appliance which is connected to a water pipe type ground to prevent the frame from becoming electrified. Use this type of wire for the ground. Consider that ground is a connection to that metal stake, conduit, water pipe or copper grounding pipe found in your home, office, apartment, etc.

As to the three-hole receptacle which has a "ground" hole to provide protection for appliances and tools, this normally is connected to the same type ground system as stated earlier through the use of a green wire or a bare wire in the box. You can use this ground connection with assurance that it will prevent trouble if your device is properly connected to the other end of the extension wire. The extension wire must have three conductors in this case.

So we find that there is much room for individual interpretation of what the code means in terms of electrical wiring. Much of the code has found a standard meaning through acceptance and practice. To get an official ruling as to what anything means, call your local electrical inspector at city hall and ask him. He is the final word on the meaning or interpretation for any particular city or area.

Being consistent is very important. Always use black for the "hot" lead. If you make a run from a switch to a fixture, use another color. If you can't do that, then continue to use black. Let *white* always be the continuous, uninterrupted common lead for all your electrical circuitry. When in doubt, call your helpful electrical inspector. If you can't reach him for any reason, then call or visit electrical shops and discuss the situation with master electricians. But don't take chances.

## LOADS AND WIRE SIZES

Branch circuits which supply household ranges, wall-mounted ovens and counter-mounted cooking units shall have a size 8 wire for ranges of 8 ¾ kw (kilowatt) or higher wattage ratings. If you have a 50 ampere branch circuit coming from the fuse or circuit breaker box, then you may be able to use a smaller size wire tap in the conductor(s). The tap must be made in a box, with approved fittings. But the conductors must be of sufficient size to carry the required current. They must have at least a 20 ampere rating, and the connections must be as short as possible to the appliance. Wire sizes for other circuits must be sufficient to carry the required current and not smaller than # 14 wire. There are some situations where a smaller wire than # 14 can be used. When you make a connection to a light fixture and the connecting wire from the fixture is not over 18 inches in length, then that 18 inch wire might be smaller than the # 14 wire which is to the box. Normally all modern homes use # 12 as the smallest wire.

## CIRCUIT BREAKERS AND FUSES

A general rule of thumb is that any circuit shall either be fused or have a circuit breaker whose size is no larger than the smallest rating of receptacles on that circuit. Normally home and shop circuits have 15 amperes in each branch. A branch is a separate line from the fuse or breaker box to that series of outlets or fixtures. One branch may have several receptacles or lights connected to it. For devices which have larger loads such as a motor, oven, air conditioner or garbage disposal, there is usually an entirely separate circuit installed just for this device. When considering circuits which have motors on them, remember that the motor will draw more current when starting then running. Thus, the fuse or circuit breaker for a motor will be larger than the others. But it should be no larger than necessary to accommodate that starting current and should always trip or blow if the motor is stopped or a short circuit develops. For up to ⅓ horsepower motors, the code suggests that a 125 percent fuse or breaker rating be used. A dishwasher which has a heating element inside it may draw, with its motor and heater, enough current (near 20 amperes) to require a separate circuit just for this unit. Fusing and breaker requirements must accommodate both the "cold" resistance of the heater element and the starting current for the motor.

Continuous loading on any branch circuit should never be more than 80 percent of that branch's current capacity. This is important if you add wires to an existing branch circuit. You must be very careful not to overload that branch circuit. Never increase the size of a breaker or fuse on an installed branch circuit when you add some circuit(s) to that branch. With everything you add to the circuit, the total circuit capability is still just what it was fused at or has a breaker for. If you need to add some kind of a high demand circuit, then *go back to the fuse or breaker box and run your new, larger feed wires from there with a new fuse or breaker*. Do *not* tap into an existing circuit for this requirement and then "overfuse" it without changing the wire size.

## CALCULATING AMPERAGE

It is well to know the simple formula:

Watts = Amperes multiplied by voltage

There are many appliances and tools which are rated in watts. You may not know what kind of a circuit they require when the circuits, fuses and circuit breakers are rated in amperes. Manipulating the above formula gives:

$$\text{Amperes} = \frac{\text{Watts}}{\text{Voltage}}$$

This is easy to solve. The voltage will always be either 110-120 or 220-240 volts. If you know the watts the device consumes, then you can determine the circuit capacity necessary (in amperes) to operate the unit. For example, if you have an air conditioning unit which requires 1800 watts, then you find:

$$\frac{1800}{110} = 16.3636 \text{ Amperes}$$

If you try to put this unit on a 15 ampere house circuit, you will blow fuses or trip breakers like mad. Then, too, the starting current of the motor will be higher than the normal running current. You might need to have a circuit with about one-third more capacity than calculated which is fused or protected by circuit breakers having this higher amperes rating.

Sometimes we continue to plug in appliances and devices on a line and never take a moment to calculate what the current capacity needed might be. Thus, we either blow fuses or trip the circuit breakers. "Hot connections" on wires may develop which damage the circuit so a fire could result.

## Amperage Meter

Sometimes we do all that is required to insure safe and correct operation of a device and still have trouble. One such example was related to me by a friend who purchased a paint sprayer with its attendant compressor. He ran in a separate line and put in a 15 ampere circuit breaker to protect the line. For the size of 15 amperes, a size 14/2 wire is sufficient. He now had an entirely separate circuit for the compressor and sprayer. When he turned on the sprayer and began to operate it, the circuit breaker tripped. After being reset, it tripped again indicating something was wrong. An examination of the relatively small unit showed that *there should* have been sufficient capacity in amperes to account for the motor starting current. Still, the unit tripped the breaker after it had been operating for a few minutes. In a case like this is it necessary to get an amperes indicator, which is a small meter that shows you how many amperes are being drawn from the electric supply. You can get one from the electric company or an electric installation company. They usually have one.

The equipment may operate all right, seemingly, if you just give it more "juice." But you know that is not supposed to be necessary. It's best to solve the problem right away and prevent big trouble later.

We had a master electrician come out to replace the fan motor on our five ton size refrigeration air conditioning unit. After he had installed the new fan, he then used an ammeter on one line to the fan. It showed a reading of 1.5 amperes. The instructions from the factory concerning the fan motor said it should not draw over 2.4 amperes. We were well satisfied that the unit was installed properly and would give no more trouble, which was a true assessment.

## Ampere Ratings of Branch Circuits

Normally the receptacles will be connected to a 15-20 ampere branch circuit. The normal rating of appliances which are connected to such circuits shall not exceed one-half of this amperage as a normal demand. In some cases there will not be any other receptacles on such circuits except those to which the appliance may be connected.

Fixed lighting units may be on 25 to 30 ampere circuits if the units are heavy duty types and the connections to the lamps are not over about three-fourths of the circuit rating. For a 30 ampere circuit, the connections to the lamps might not have a rating over about 24 amperes. This is a safety factor.

The higher amperage circuits in the range of 40 to 50 amperes for stoves, clothes dryers and heaters will have one circuit just for each device with nothing else on that circuit. It has been recommended that clothes dryers have at least a 40 ampere circuit. A water heater and any other heavy duty electrical devices must have a circuit based on their wattage consumption.

## INSTALLING RECEPTACLES

If you are building a home or renovating an old one, it is well to know that the code says no point on any wall shall be over 6 feet from a receptacle. This may be measured at the floor line. If you have a den and you want to be in accord with this code, then you may place two receptacles 12 feet apart. You will have them such that no point on the floor between them is more than 6 feet from either one. If there is a doorway in a wall, you should have a receptacle within 6 feet of the doorway. For devices such as clothes washers and dish washers, you need a receptacle within 6 feet of the planned location for these units. In pactice, however, the outlets for these devices are usually much closer, say within 3 or 4 feet of the planned location.

Sometimes in older homes you have walls in which you cannot put receptacles as you do with most modern homes. In this case you put them in the floor, close to the wall if possible. The code says that if you are within a foot or so of the wall, then the number of outlets required (or suggested) is the same as for in the wall. Install a receptacle every 6 feet or so, no more than 12 feet apart. You may have them closer together for certain kinds of lamps or things you plan to operate electrically. And you may have a multitude of circuits needed if you have lots of outlets. Remember these circuits should be the 15-20 ampere type, and you must plan for not over 50 percent constant use of the available electrical power on one circuit.

In kitchen areas over countertops, it is stated by the code that every place where the counter-top is over 12 inches wide shall have a small appliance receptacle installed. These may be such that a 5 foot cord can easily reach them from any place on the counter-top. Of course, with small movable appliances, one such outlet with two sockets is usually sufficient for each counter top.

It is always suggested that one outlet be installed in the bathroom near the basin so you can use devices like shavers there. Remember, though, that there should be a ground fault interrupter on these circuits especially in the bathroom. In single-family dwellings it is a requirement that there be at least one outside outlet of a proper weatherproof type available. Sometimes this requirement is met by having a couple of light outlets on the back and front doors of the house. You can remove the light and put in a plug socket in the lamp socket so you can attach a yard tool run by electricity. This is done in many homes. Be careful that you don't overload that circuit.

## LIGHTING SYSTEMS IN HOMES

The code says that there must be at least one light with an "on and off" switch in every room, hallways, bathroom garages and at outside entrances. The code goes further to say that you should have a light in any attic, utility room, basement, or other area underneath the house proper. It is always a convenience to have a light in any room or area, such as underneath the stairs, where you might have planned to store items.

Of course, you are familiar with the outlet which can be swtiched on and off. The code says that if you do not want an overhead light in some room or area, an outlet into which a light can be connected will be adequate. Then, too, you can have automatic control of lights in areas like the backyard, garage and entrances which can be controlled by an automatic type of switch like a set timer. There are devices which turn the lights on and off automatically when it gets dark and light. Yard lights may use this kind of control element. We always like to have a switch in the circuit so that, at our option, we can control the light or perhaps some device plugged into that light socket. Then we can add the daylight-dark automatic control type unit.

### Kitchen Area Light Fixture

In Fig. 2-3 we show one type of light fixture which was installed in the kitchen-breakfast room area of a modern home. You can see a section of wire which is used to keep the fixture high while work is performed underneath it.

The large bulb, tends to diffuse the light and keep it from glaring. The fixture is large enough so that a large size light can be used, up to 250 watts. There will be plenty of light from this one fixutre for the general kitchen-breakfast room area. Of course, there will be concealed lights over the sink and stove to supplement this light. This fixture itself is attractive in its setting and adds to the decor of the room area. These kinds of considerations should always be taken into account in the planning or modification of lighting for any home.

### Living or Dining Room Light Fixture

In Fig. 2-4 we show a lighting fixture for the living room-or dining room area. Notice that it is more formal and can use smaller lights to give the intimate kind of atmosphere which is usually

Fig. 2-3. A kitchen-breakfast room light fixture.

73

Fig. 2-4. A living room-dining room light fixture.

desired in these areas. The home here is so new that the wallboard has just been put in place and the walls are not yet painted or papered. Sometimes the fixtures are hung at this stage of development, and sometimes the contractor waits until the walls and floors are finished.

## Dimmer Switch

It is nice at times to have such a fixture controlled by a "dimmer switch" so that the light level can be precisely adjusted. If

74

you use the living room for games or gatherings where lots of light is needed, this can be had at the turn of a dial. For conversation you can turn the light down to an intimate level which is always acceptable. You *can* easily replace a standard switch with a dimmer switch.

### Safety Plugs

The wall receptacles in one home were carefully arranged to provide safety against a child inserting something into the receptacle and causing an accident. As Fig. 2-5 shows, the snap-in plastic safety plugs are used on all receptacles and sockets which aren't used for lamps or devices. Safety plugs like these can be purchased in stores where they sell electrical supplies. We highly recommend them for a baby's or child's room. By the way, notice the exposed three-hole socket. The center lower hole is the grounding connection we've discussed.

### WIRING FOR CLOTHES DRYERS, REFRIGERATORS AND SIMILAR APPLICANCES

Dryers, refrigerators and other appliances probably will not be present at the site when the electrical wiring is installed. There must be some way to consider what size wires are needed to service the clothes dryer receptacle or the box where it will be connected. The code says that you should have a service which can handle 5000 watts. Remembering our previous formula:

Fig. 2-5. Use the plastic, snap-in safety plugs.

$$\frac{5000 \text{ watts}}{110 \text{ volts}} = 45.4545 \text{ amperes}$$

If the voltage is to be from the 220 volts supply:

$$\frac{5000 \text{ W}}{220 \text{ V}} = 22.7272 \text{ amperes}$$

Let us see what this means in terms of wire size. Referring to Table 2-1, we find that a size 8 (8/2) wire will be needed at the 110 volt level. A size 14 (14/2) will do at the 220 level. Actually since a dryer will have a motor, a size 10/2 will probably be used because of a motor's high current demand when starting.

The code says you must provide for this 5000 watt load or know specifically what size dryer will be used in the home. The code also says that usually the inspectors will require a size 8 and that the branch circuit should not be loaded to more than 80 percent of its capacity. You find the amperage for 80 percent of, say 40 amperes, by multiplying the 40 by 0.8. That gives you 32 amperes. You would use a current meter clipped around one line of the dryer to see what current the machine "pulls" when it is working and to check that it is not overloading the wire capacity.

### Wattage Check

Now the previous section brings up some interesting considerations. In time your kitchen appliances will wear out and have to be replaced. How do you make sure the new appliance will not overload the wiring which was put in for the old unit? You check the wattage requirement of the new unit against that of the old unit. Few people do this. It is up to you to make sure the circuit can handle the new units by making the wattage check.

If you find, for example, that a new dryer unit will overload your dryer circuit, then you need a new circuit installed with wires which can handle the new load. This is seldom needed in homes which have been planned well. In these homes the circuits are fully adequate to handle, usually, more load than is planed on them initially. If you plan a home, be sure you do have adequate circuit capacity for expansion and growth. You *never* simply replace the existing size circuit breaker (or fuse) with a larger one simply to accommodate a larger size washer or dryer. The circuit may work, but don't do it unless you know, by having it checked by a reputable master electrician, that you have an adequate circuit wiring capability.

Normally you won't find any circuit fused or protected by a circuit breaker which is much smaller than needed for that circuit. In 99 percent of the cases, using a larger circuit breaker requires a new pair of wires to be run to the device location. Such wiring then can handle, easily, the increased amperage load.

### Surface Conduit

Running the wires to the outlet from the circuit breaker panel may not be an easy task. If the original wires were of the plastic sheathed type, there is a good probability that this cable has been stapled or fastened along its length inside the wall. You cannot just pull the wires out and run some new and larger wires in place of the old ones. What you have to do is make a new run, probably from the panel to the outlet. This can be done in a number of ways. You might use surface conduit which is a rectangular kind of metal made in two parts, a base and a top, that snaps into place. See Figs. 2-6A and 2-6B.

The inside cross section of the conduit can be from ½ × ¾ inches, to whatever size in inches is necessary. This kind of conduit fastens to the baseboard or along the ceiling or walls. It can be made pretty inconspicuous by painting or by getting it in a color which matches the wallpaper or painted wall. The exit from the wall interior to the panel box will have to be through a hole in the wall at an appropriate place near the panel. You will have to use the proper fittings for the situation. This kind of conduit is available from most hardware and electrical supply businesses.

### Tubular Conduit

In some cases you can use tubular conduit which runs around the home on the outside (Fig. 2-6B). This starts at the panel and comes to the point on the wall where a hole permits entrance to the location for the appliance or light receptacle. This kind of conduit can be installed such that it is waterproof and so fastened that it is permanent. With this type of installation, you do not see any evidence of this added wiring inside the home.

### Fishline

The outside conduit usually requires a master electrician's installation. You first fasten the conduit in place with clamps, and then pull the wires through it. A so-called "fishline," which is a thin, long and flexible metal strap, is used to "fish" the wires

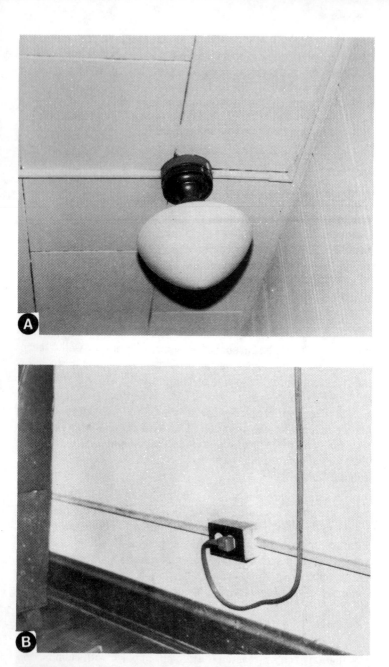

Fig. 2-6. Installation of surface conduit. (A) Surface conduit installed for a light and a receptacle. (B) Conduit from the box into the basement. (C) Clamps fasten the conduit to the brick wall.

through the conduit. Also, it will be necessary to use "pull boxes." These are simply *terminations of the conduit* into which access to the conduit can be had in order to shorten the "pull" runs of the wires, if the distance from the panel is long. Going around corners requires special corner fittings. You must be able to open them to pull the wires to the corner and then insert them into the next run from that junction.

## Checking for Conduit Connections

If your home has Greenfield conduit—the hollow armored conduit into which wires can be pulled—then you may be able to run new wires in place of the old ones which must be pulled out. This is also true for rigid conduit which is often used in some homes, shops and offices. The rigid tube conduit is more often found in business establishments than in homes. Check the attic to see how the wires come to the boxes.

If you have no attic or cannot investigate how the wires come to the boxes, you might try examining some of the boxes to see if they show a conduit connection or not. This is not a really good way to find out if conduit is used throughout the walls or crawl spaces, etc., of your home. Sometimes metal fittings are used on plastic sheathed cable and, from the box inside view, it looks like a metal pipe or armored flexible cable (BX type) has been used. You could get a wrong idea, but you might look. If you find any with plastic sheathed cable entering the box, that is a good sign that metal conduit probably has not been used in your installation except, possibly, for very short runs to some appliances or heavy duty light fixtures.

## WIRING ON THE OUTSIDE OF BUILDINGS

The code says that if you use 600 volts or less, you can use the following methods of wiring.

■ Support the conductors on insulated supports.

■ Use a multiple conductor cable approved for outside use.

■ Use an aluminum sheathed cable (ALS).

■ Use the rigid conduit.

■ Use a "busway" which is a metal type trough which meets the weathering requirements and specifications for rigidity.

■ Use of metallic electrical tubing such as the flexible armored cable we discussed. However, be careful as this kind of cable is not waterproof. Get an approved type for use outside.

■ Use an *intermediate* metal conduit, and you have to get a definition from the city inspector as to what this means. The code says it is a metal raceway of circular cross section with all fittings, connectors, etc.

The code also says that if you use a branch circuit to supply current-voltage to lights or lighting fixtures mounted on the outside of buildings or on poles, the voltage shall not be greater than 150 volts to ground. There are some exceptions, but this rule covers that wiring which is not to exceed 150 volts. You must plan to use 110-120 volts as the normal supply.

## CONNECTING BRANCH CIRCUITS TO BUILDINGS

Unless you use underground wiring to make the connection from the home electrical system to the pole or whatever has the light on it, then you must have the wiring a specific number of feet above the ground as follows. It must be at least 10 feet high and, if over a driveway, must be at least 18 feet high. If it is a residential driveway, then the clearance must be at least 12 feet. So do not run wiring from the house to the garage, for example, which is 6 or 8 feet from the ground. Make sure it is at least 10 to 12 feet high. You can use an underground feed system with appropriate conduit or sheathed or protected cable which is designed for the underground type of installation. Then bury it deep enough so that normal digging, etc., won't produce a cut in the cable or damage it.

You have to be careful especially when running electrical conduit or metallic covered cables in the earth. According to the code, you cannot use these where there may be any kind of chemical action which might corrode or destroy the metal. Avoid cinder fills or any area where acids may be present. Of course, the metallic conduit or covering must not be subject to water damage. If you plan to use underground wiring, then be sure you get a cable and wires which are approved for this purpose.

## WIRING UNDER PLASTER

If it is necessary to make an electrical wiring extension under plaster, then you *must* use either rigid conduit, the armored type flexible conduit (BX type), type AC cable or a type MI cable. You can run the wires in an approved trough or raceway. If you ask what that "under plaster" type of installation amounts to, consider that it is a wiring scheme which is fastened to masonry or whatever kind of wall and then plastered over when the wall finishing is performed.

It is important to realize that the code says you cannot go from one floor to another unless you install rigid metal conduit to carry the wiring. Or you can use the AC or MI type cables.

## USING PLASTIC CONDUIT

Some types of rigid nonmetallic conduit (plastic pipe) in which to draw the electrical wires has been approved for installation. One type which is approved for underground installation is high density *polyethylene*. A type which is commonly used in above ground installations is the rigid *polyvinyl chloride* (PVC). Before using any such "pipe," it will be a good idea to check with the city inspector. At least see if it has the Underwriters Laboratories (UL) label saying it is approved for use in the type of location in which you plan to install it. Of course, it must be so installed that it cannot be broken or harmed in any way. If you plan on going through concrete, such as for the wiring in a subfloor of a building, then rigid conduit of the metal type approved for this purpose is required. You cannot use plastic pipe for this purpose.

You need a cable deep enough so frost won't get to it. Make sure the area is not damp or that there is no likelihood of digging or earth change movement to cause damage. You might use a cable which is moisture-resistant, moistureproof. A moistureproof installation would be a conduit which is sealed and airtight as well as waterproof. The cable types which are suitable for use in the damp locations according to the code are RHW, TW or THW, RUW, ALS, MI cable, and, of course, the lead-covered cables. Some of these may be approved for direct burial in the earth, such as the lead-covered cable, the ALS and the MI cable. But it is always wise to check with the local city inspector to see what is approved by the city ordinances. In Fig. 2-7 you can see how a flexible plastic pipe with a cable inserted inside it is run through the ground to make connection from the house to a garage. Notice the fittings which make it watertight.

## LOCATION OF OUTLETS

We have stated earlier that there should be, according to the code, a receptacle on each wall so that a 6 foot cord to a lamp or other device can be plugged into it. This means that along a wall the receptacles should not be over 12 feet apart. If the wall is broken by a doorway or two windows, then the receptacles should be under them at least, or between them if doorways are involved.

Fig. 2-7. Fittings on an underground electrical wiring installation from a house to the garage.

Now we consider the height of the receptacles and switches from the floor. We have found that in most modern homes the receptacles are placed on an average of 15 inches from the floor to the center of the receptacle. Wall switches are located 4½ feet above the floor. On countertops where there are cabinets above and a counter below, such as in a kitchen, the receptacles and

**Table 2-2. Number of Conductors in Deep Boxes.**

| Box Size | Maximum Number of Conductors of size: | | | |
|---|---|---|---|---|
| | #14 | #12 | #10 | #8 |
| 3¼ × 1½· | 5 | 4 | 4 | 3 |
| 4 × 1½· | 8 | 7 | 6 | 5 |
| 4 × 2½· | 11 | 10 | 9 | 7 |
| 4.75 × 1½· | 16 | 14 | 12 | 10 |
| 4.75 × 2.12 | 23 | 20 | 18 | 15 |
| 3 × 2 × 1½· | 3 | 3 | 3 | 2 |
| 3 × 2 × 2 | 5 | 4 | 4 | 3 |
| 3 × 2 × 2½· | 6 | 5 | 5 | 4 |
| 4 × 2.12 × 1½· | 5 | 4 | 4 | 3 |
| 4 × 2.12 × 2.12 | 7 | 5 | 6 | 5 |

(The three dimension sizes are switch boxes, etc.)

switches are usually located symmetrically, halfway between the two. In bathrooms the wall receptacles near the basin are located from 14 to 16 inches above the basin level if they are over the lavatory counter. If they are located in a wall away from the counter, they may be slightly lower.

Sometimes in homes we find outlets which are placed more than 5 feet above the floor. The code says these do not satisfy the requirements for outlets in rooms which must be placed so that a 6 foot cord, measured along the floor, can reach one of them. So, if you plan outlets at 5 or more feet above the floor or want to have them in cabinets or cupboards, you should have the other receptacles in this area placed either above countertops or near the floor level. They could be at the common level of 15 inches from the floor.

## ELECTRICAL SPACE HEATING

The code says electrical space heating will be governed by what the local inspection authority might require for approval. In the times of less oil and other such means of heating, it is not inconceivable that perhaps there will be more use of electrically operated space heaters. These may range from types which can be installed in walls at baseboard levels to individual heaters which, like stoves, may be placed in various rooms. Depending upon the size of the electric heating unit (like an electric dryer), you may be required to have separate branch circuits installed. Have these connected to individual circuit breakers at the electric breaker panel. Calculations for the size wire can be made from the wattage produced by these heaters. Also, you can determine what the loading effect and requirement will be on branch circuits already existing in your home, when you plug in the portable type electric heaters.

You must have expert advice by a master electrician if you plan on permanently installed space heaters due to the chance of fire danger if they are not properly installed. Remember that in an existing home, you might have many outlets on a given branch circuit. The limit for them totally is 15 amperes usually, using 14/2 wire. Don't make the mistake of thinking that if you unplug a large lamp from one receptacle in a room, the branch circuit which feeds that receptacle is absolutely free of any other demand. Unless you know which receptacles and lights are on a given branch, be very careful about what size electric heater you plug into an existing outlet so that you won't overload that circuit.

Of course, if it is overloaded, the circuit breaker should trip, or a fuse should blow out to give you protection before a dangerous fire level is reached. But it is wise to try to keep your electrical loads reasonable and below the circuit breaker protection point. Most portable electric heaters and other devices have a wattage rating on the name plate or box when you buy it. Look at it and see how the rating compares to, say, the use of a light which might have a 250 watt bulb or several lamps which might be used in the room. That gives you some idea of current demand.

You can always calculate the amperage requirement because you know the voltage (110-120V) and the wattage. You can see how much of the normal 15 ampere circuit capacity you will be using with this device, whatever it is. Consult with a master electrician who can inspect circuits and advise you whether you need a "new run" for your projects.

### Table 2-3. Number of Conductors in Shallow Boxes Less Than 1½ Inches Deep.

| Box Size | Maximum Number of Conductors of Size | | |
| | #14 | #12 | #10 |
| --- | --- | --- | --- |
| 3¼ | 4 | 4 | 3 |
| 4 | 6 | 6 | 4 |
| 4.75 | 8 | 6 | 6 |
| 1¼ × 4 | 9 | 7 | 6 |

The fittings must be taken into account if they
project inside the box. Also, when using cable (Romex)
estimate the number of conductors by comparing one conductor
of, say, size 12 to the diameter or cross section of the cable.
If that cross section equals 4 #12 sizes, then the cable equals 4 wires.

**Table 2-4. Number of Conductors Permitted in Sizes of Conduit.**

| Wire Size | Conduit Diameter in inches | | | | | |
|---|---|---|---|---|---|---|
| | ½ | ¾ | 1 | 1¼ | 1½ | 2 |
| 18 | 7 | 12 | 20 | 35 | 49 | 80 |
| 16 | 6 | 10 | 17 | 30 | 41 | 68 |
| 14 | 4 | 6 | 10 | 18 | 25 | 41 |
| 12 | 3 | 5 | 8 | 15 | 21 | 34 |
| 10 | 1 | 4 | 7 | 12 | 17 | 29 |

## SUMMARY

The National Electrical Code is published by the National Fire Protection Association. You may want to obtain a copy of the code as it is specifically written by the technical committee of that association. Your local fire station chief or the city electrical inspector can probably assist you in getting the information you might need to write to this association. Be aware that there may be lots of publications which contain information based on the code, but they are *interpretations* of information contained in the code. We have made eveto be accurate and to analyze and interpret correct if there is any doubt in you mind or you want more information, get a copy of the code itself. Write to the National Fire Protection Association, 470 Atlantic Avenue, Boston, MA 02210. Consult Table 2-2 through 2-4 for information on the numbers of conductors in boxes and conduits.

If you are planning a new home or renovating an older home, you might want to look pretty far into the future and install circuits (or have them installed) which can handle a larger ampere load than the code required. For example, there is no reason you cannot have 12/2 wire used instead of the smaller size 14/2. You might even have circuits which have 10/2 wires. Planning for the future use of electricity in any kind of home will simply mean planning for increased use of electricity. There will be more and more devices which depend on electricity for their operation. So be generous to yourself. Plan plenty of outlets and circuit capacity for additional and special lighting, and plan circuits which can stand an increased amperage demand.

# Electric Company Requirements, Services and Installations

**3**

Now we will consider some requirements of an electric company regarding what must be done to get electric power into your home, office, condominium, etc. The requirements are general in nature so that they can apply in whatever area of the country you live. We will be considering the requirements for new homes concerning the metering and service inlets. If you have an older home or want to modernize, you may have to change the service input wire size and panel, etc., before the electric company will connect you to its supply lines. It is important for you to know where the electric company's responsibility ends and yours begins, with regard to the service input.

## LEARNING THE HARD WAY

Let's relate a personal story. Our own service input installation is an underground feed from a pad-mounted transformer to the meter on the home's side and then to the breaker panel. We found that the pipe going from the meter down to the ground had rusted and corroded through. See Fig. 3-1. The wires were exposed at ground level.

Under the assumption that "everything from the meter on" would be the property of, and serviced by, the electric company, we called them. They sent a service representative out to examine the problem. We were told that the company responsibility ended with the connection to the *transformer*. Also, we had to hire a

Fig. 3-1. A pad-mounted transformer for a new home with its connecting wires.

qualified master electrician to come out and replace the conduit before problems developed due to the exposure of the wiring at the service input . We were told, however, that the wiring was of the special type which was waterproof and weather proof. It could be run under ground without damage to the conductors themselves in case the conduit around the wires failed in some manner.

Notice in Fig. 3-2 how the light company in this type of installation has run the "high lines" along poles to the vicinity of the home. Then a conduit was used to run the wires down into the ground and to the pad-mounted transformer in Fig. 3-1. In this particular installation the lines to the house meter were run through approved plastic pipe from the transformer to the meter. They were also underground. Figure 3-3 shows how the lines appear coming up through their conduit to the vicinity of the breaker panel, and also shows the wires coming out of the house which must be connected to the breakers in the panel. This location of where the wires come out of the house is called the "home" location of wiring by professionals. A wire which comes from some receptacle and runs directly to this location is said to go from the receptacle on a "home run."

Let's resume our story. When we found that we had the responsibility and the expense of running new wires and conduit from the transformer to the meter, we learned something the hard way. For what it is worth, we pass along this information to you. *You* have the responsibility for the service input to the home,

Fig. 3-2. The "high line" connection to the pad-mounted transformer. The pad is a concrete base.

including such wiring as may come from a transformer owned by the electric company. Digging up underground cables and conduit and replacing same is not an economical venture. Perhaps you will want to ask your own electric company's representative to tell you specifically where your responsibility begins and where theirs ends on your electric service input. At least it will prevent surprises later on if it is necessary to change, repair or alter your service input to your home or office. If you live in an apartment you may have no responsibility in this area. That is probably the case, unless you are leasing. It might pay you to ask about it anyway.

## GUIDELINES FROM THE ELECTRIC COMPANY

Through the courtesy of the Houston Lighting and Power Company we are able to state and discuss some important electrical installation guidelines which they furnish to their customers and to contractors, architects and engineers who need and use this kind of information. It will pay you to visit your own electric company and ask for their guidelines, which are intended to help you gain the greatest benefit from the use of electricity in your home, apartment, shop, business, etc.

The following guidelines are intended to supplement and not replace or be in conflict with the current editions of the National Electrical Code, the National Electrical Safety Code or any regulatory authority having jurisdiction. We now examine those guidelines, including only those which are general in nature and of benefit to all of us.

## SOME DEFINITIONS

■ **Application for Service**. The agreement or contract between the electric company and the customer under which electric service is supplied and taken.

■ **Connected Load**. The combined manufacturer's rated capacity of all motors and other electric energy consuming devices on the customer's premises, which may, at the will of the customer, be operated with the electric energy to be supplied from the service of the electric company.

■ **Customer**. Any individual, partnership, association, firm, public or private enterprise, corporation or governmental agency now being served, or hereafter being served, electrical energy by the electric company at any specified location.

■ **Customer Extension**. Any addition to the company's existing distribution system required to be made in order to render electric service to a customer.

■ **Customer's Installation**. All wiring, appliances or apparatus of any kind owned or operated by the customer on the customer's side of the point of delivery (except the company's metering equipment) used in connection with the customer's ability to take and use electric service of the company.

■ **Point Of Delivery**. The point where the electric energy first leaves the line or apparatus owned by the electric company and enters the line or apparatus owned by the customer, unless otherwise specified in the customer's aggreement for service. This is not necessarily the point of location of the electric company's meter.

■ **Customer's Service Equipment**. The necessary equipment and accessories, located near the point of entrance of supply conductors to a building, which constitutes the main control and means of disconnecting the supply to that building. This equipment usually consists of a circuit breaker or switch and fuses.

■ **Distribution Lines**. The company's lines located along streets, alleys, highways or easements on private property when used or intended for general distribution of electric service to customers of the electric company.

■ **Electric Service**. The availability of electric power and energy, regardless of whether electric power and energy is actually used. The supplying of electric service by the electric company consists of the maintaining by it, at the point of delivery, of approximately the agreed voltage, phase and frequency, by means of facilities adequate for carrying the load which the electric company is obligated to supply. Notice here that there seems to be no provision for the supply of sufficient electrical energy for future requirements. This could mean that if your demand does increase, then the electric company would have to put in perhaps larger transformers, etc., to supply the new demand. A new agreement with them might also be required. It will be well worth the effort to check this out if you plan expansions of electric service demands for any reason.

■ **Meter**. An instrument, or instruments, together with auxiliary equipment, for measuring the electric power and energy supplied to a customer.

■ **Meter Mounting Devices**. The device owned and furnished by the electric company for mounting and/or enclosing the company's metering equipment.

■ **Meter Socket Base**. A special meter base for outdoor and indoor use, furnished by the electric company and installed by the customer at his expense.

■ **Meter Box**. A metal cabinet for outdoor use installed by the company. The board for this mounting is installed by the company. A metering transformer box is another type of metal cabinet furnished by the electric company, but installed at the customer's expense for enclosing the company's metering current transformers.

■ **Meter Loop**. The opening in and extension of the customer's service entrance conductors provided for the installation of the company's meter.

■ **Rate Schedule Classification**. The classification of the customer's electric service, the amount of power supplied, the rate area and the purpose for which the electric service is to be used.

■ **Service Entrance Conductors**. The wires or bus bars provided by the customer extending from the customer's service equipment to the terminals of the service drop or service lateral.

■ **Service Lateral**. The underground service conductors between the street and/or easement main, including any risers at a pole or from a transformer, and the first point of connection to the service entrance conductors in a terminal box or meter. Underground service conductors are owned and maintained by the Customer.

■ **Service Outlet**. The outside terminal portion of the customer's installation to which the electric company's service drop is connected.

■ **Service Drop**. The overhead service conductors extending from the electric company's overhead distribution system to the customer's service entrance conductors at the point of delivery.

■ **Service Outlet Location And Data Statement**. A written statement prepared by the distribution engineering division of the electric company for the guidance of the company and the customer or his representative. This statement shows the estimated load to be served, the type of service which the company proposes to make available, and the agreed location for the customer's service outlet at the specified premises under consideration.

■ **Type Of Service**. The characteristics of electric service as described in terms of frequency, phase, nominal system, voltage and number of wires.

■ **Underground Distribution Areas**. Those areas where electric service is supplied by the electric company from its underground distribution facilities. Underground street network areas are designated by the company where established 120-208 volt street secondary network systems are in operation. Underground radial areas are those areas where electric service is supplied by the company from its underground distribution facilities connected to a radial supply. Underground residential distribution areas are those areas where special contractual arrangements have been made for single phase 120-240 volt underground service to all customers in a subdivision or specified area.

## KNOW YOUR RESPONSIBILITIES

In some respects these definitions explain definitely just what your responsibilities as a customer are with respect to the type of service you plan to use from the electric company. It is true that most persons will not be too concerned with the electric service requirements as they will have been met in building the house; the contractor has that responsibility. In expansions, modifications and maintenance of service facilities, it is very important that you know what you are accountable for and must provide. If your demand for electricity increases for whatever reason, it may be necessary to go clear back to the electric company's "drip" or underground feed and have that changed to a larger size to adequately meet this increased demand.

## OVERHEAD AND UNDERGROUND SERVICE AND SPECIFICATIONS

While we are on the subject of service feeds from the electric company, it will pay us to examine some specifications for these type services as shown in Fig. 3-3, 3-4A and 3-4B. Recall that your responsibility is from the point of attachment of the electric company's lines to your home or residence. It has nothing to do with that "stuff" which is connected on the "non-house" side of your meter. You will do well to go back and review the point of attachment of the electric service if you are not familiar with this concept.

In Fig. 3-5 we see a similar but perhaps more specific example of the underground installation. In the center you note the feed from the electric company's pole to the transformer (pad mounted or in a vault underground) and then to the home or residence. At the top of Fig. 3-5 are detailed considerations for other services such as gas and telephone which may have lines in the public easement. At the bottom of Fig. 3-5 is shown the arrangement of the meter, the circuit breaker panel which connects all lines inside the home to the electric company's feed lines, and that all-important ground connection.

## Meter Socket

The meter socket shall be furnished by the electric company and installed by the customer's electrical contractor (Fig. 3-6). The location should be on the side wherever practical, but may be at any point around the perimeter of the house or garage where accessibility, height working clearance, etc., comply with the service standards.

## Service Cable

The customer's contractor shall furnish and install service to the following specifications.

—Conductors shall be type use and sized for the load according to the National Electrical Code. Conductors shall be clearly marked as to type use.

—Cable shall be installed a minimum of 2 feet below grade. Backfill in the bottom of the trench and above the cable shall be fine sand or soil and free of rocks, concrete or hard objects which might damage the cable. Aluminum and aluminum alloy cables require the utmost care in handling and installation. Most installations are especially susceptible to nicks and scratches. Careless handling may result in failure of the cable.

The customer's contractor shall install service cable to within 1 foot of the secondary service hole or transformer pad. See Fig. 3-1. Contractors may contact the electric company for the proper location of the service cable connections, even if they seem apparently evident on the grounds. Service brought to the transformer pads shall be left opposite the small painted "V" mark on the concrete base. Ten feet of service cable shall be left for connection to the transformer, 5 feet of service cable shall be left for connection to a secondary pedestal. The cable shall be coiled and secured, clear of the ground, to a stake opposite the "V" mark. Cut ends shall be made watertight by an approved sealing method immediately after cutting. Caution should be observed when digging in the area (easement) to avoid damage to telephone and other cables and gas pipe connections or coatings. Damage to any utility equipment shall be immediately reported to that utility. The electric company does not assume responsibility for damage to any lines such as these by anyone other than their own personnel.

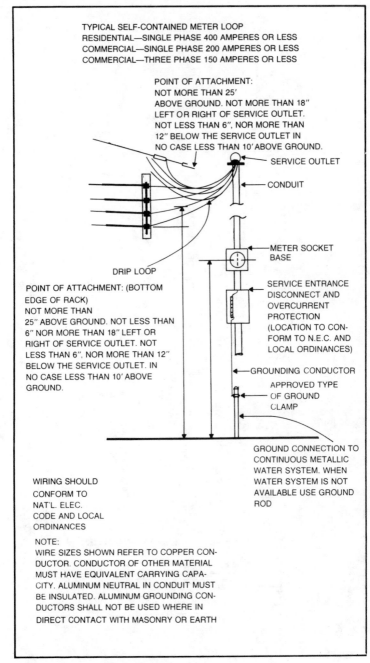

TYPICAL SELF-CONTAINED METER LOOP
RESIDENTIAL—SINGLE PHASE 400 AMPERES OR LESS
COMMERCIAL—SINGLE PHASE 200 AMPERES OR LESS
COMMERCIAL—THREE PHASE 150 AMPERES OR LESS

POINT OF ATTACHMENT:
NOT MORE THAN 25'
ABOVE GROUND. NOT MORE THAN 18"
LEFT OR RIGHT OF SERVICE OUTLET.
NOT LESS THAN 6", NOR MORE THAN
12" BELOW THE SERVICE OUTLET IN
NO CASE LESS THAN 10' ABOVE GROUND.

SERVICE OUTLET

CONDUIT

METER SOCKET
BASE

DRIP LOOP

POINT OF ATTACHMENT: (BOTTOM
EDGE OF RACK)
NOT MORE THAN
25" ABOVE GROUND. NOT LESS THAN
6" NOR MORE THAN 18" LEFT OR
RIGHT OF SERVICE OUTLET. NOT
LESS THAN 6", NOR MORE THAN 12"
BELOW THE SERVICE OUTLET. IN
NO CASE LESS THAN 10' ABOVE
GROUND.

SERVICE ENTRANCE
DISCONNECT AND
OVERCURRENT
PROTECTION
(LOCATION TO CON-
FORM TO N.E.C. AND
LOCAL ORDINANCES)

GROUNDING CONDUCTOR

APPROVED TYPE
OF GROUND
CLAMP

GROUND CONNECTION TO
CONTINUOUS METALLIC
WATER SYSTEM. WHEN
WATER SYSTEM IS NOT
AVAILABLE USE GROUND
ROD

WIRING SHOULD
CONFORM TO
NAT'L. ELEC.
CODE AND LOCAL
ORDINANCES

NOTE:
WIRE SIZES SHOWN REFER TO COPPER CON-
DUCTOR. CONDUCTOR OF OTHER MATERIAL
MUST HAVE EQUIVALENT CARRYING CAPA-
CITY. ALUMINUM NEUTRAL IN CONDUIT MUST
BE INSULATED. ALUMINUM GROUNDING CON-
DUCTORS SHALL NOT BE USED WHERE IN
DIRECT CONTACT WITH MASONRY OR EARTH

Fig. 3-3. Overhead feed and specifications.

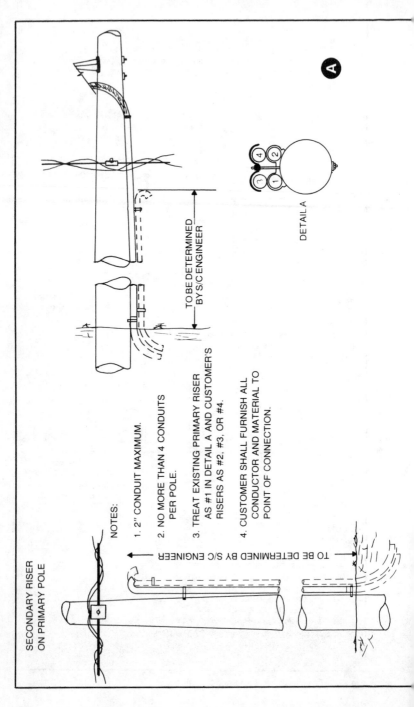

SECONDARY RISER
ON PRIMARY POLE

NOTES:

1. 2" CONDUIT MAXIMUM.

2. NO MORE THAN 4 CONDUITS PER POLE.

3. TREAT EXISTING PRIMARY RISER AS #1 IN DETAIL A AND CUSTOMER'S RISERS AS #2, #3, OR #4.

4. CUSTOMER SHALL FURNISH ALL CONDUCTOR AND MATERIAL TO POINT OF CONNECTION.

TO BE DETERMINED BY S/C ENGINEER

DETAIL A

A

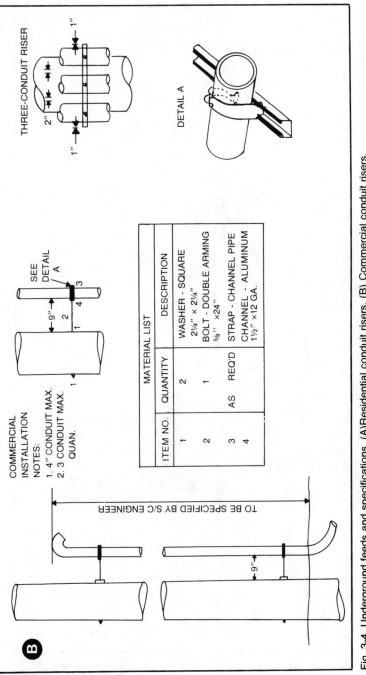

THREE-CONDUIT RISER

DETAIL A

SEE
DETAIL
A

COMMERCIAL
INSTALLATION
NOTES:
1. 4" CONDUIT MAX.
2. 3 CONDUIT MAX.
QUAN.

| MATERIAL LIST | | |
|---|---|---|
| ITEM NO. | QUANTITY | DESCRIPTION |
| 1 | 2 | WASHER - SQUARE 2¼" × 2¼" |
| 2 | 1 | BOLT - DOUBLE ARMING ⅝" ×24" |
| 3 | AS REQ'D | STRAP - CHANNEL PIPE |
| 4 | | CHANNEL - ALUMINUM 1½" ×12 GA. |

TO BE SPECIFIED BY S/C ENGINEER

**B**

Fig. 3-4. Underground feeds and specifications. (A)Residential conduit risers. (B) Commercial conduit risers.

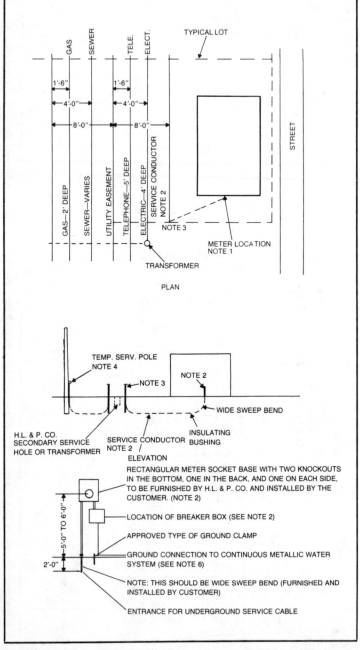

Fig. 3-5. Detailed underground installation and feed to home or residence.

Fig. 3-6. Meter socket and circuit breaker panel.

Fig. 3-7. A temporary electrical service for construction personnel comes down from the pole transformer.

## Service Poles

Temporary services poles shall be set outside the confines of the easement. Otherwise, they should be installed according to previous specifications.

For temporary service only, the height of the meter socket on temporary service poles in urban residental areas (URD) may be reduced below the requirements. The height of the meter socket shall be such that the opening for the meter is from 5 to 6 feet above ground if so required by the local building codes. The meter socket may not be mounted on a stairway and must be at a location where it is easy to read. Figure 3-7 shows one installation where a temporary electrical feed has been provided in an area where building was going on. The workmen could use electrical saws, hammers, etc.

## Proper Grounding

All services shall be properly grounded. The National Electrical Code requires grounding to a metallic underground water piping system if available. Acceptable alternatives include a driven ground rod. Regardless of the type of grounding electrode used, the electrical code requires that the metallic water system (cold water pipes) be bonded to it. In Fig. 3-8 you can see how a copper line in the home under construction has a grounding cable con-

Fig. 3-8. Connection to a copper water line for grounding.

101

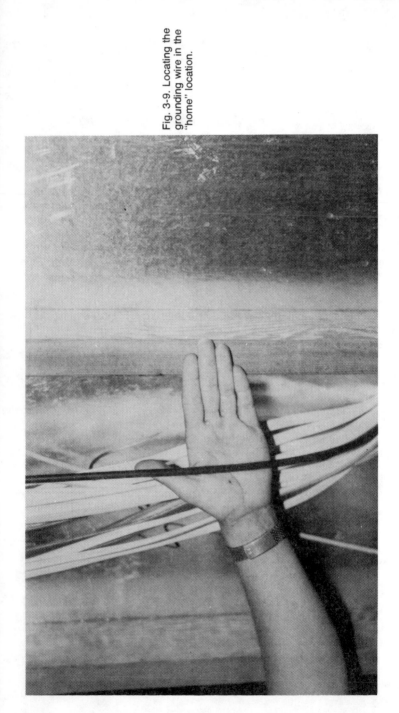

Fig. 3-9. Locating the grounding wire in the "home" location.

nected to it. This cable is then run across the building so that it can be fed outside at the "home location" and then connected to the grounding rod just outside and below the meter and circuit breaker panel. In Fig. 3-9 the workman is holding the isolated grounding cable in his hand. Notice how it differs in color and size from the other "home run" conductors which go to the circuit breaker panel.

## Breaker Box

The breaker box may be installed inside or outside the building. Figure 3-10 shows a breaker box installed in a small apartment under construction. The location is in an inconspicuous hall closet, although you cannot see that this is so. In apartments and condominiums this type of location is usually the case. In home and residential areas and buildings the circuit breaker box is usually located outside on a wall of the building or garage near the meter. You will notice the neat and professional wiring inside the circuit breaker box.

## Needs of Rural Residents

Construction in areas where electrical installations are governed by city ordinances shall meet requirements of all applicable ordinances and codes. We need to point out that some of our

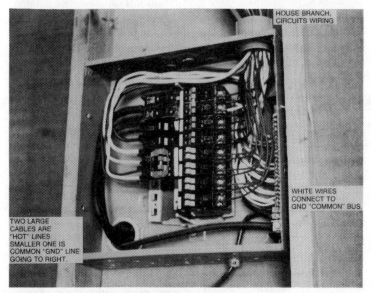

Fig. 3-10. Circuit breaker box in a home or apartment under construction.

readers may live in areas which are outside city limits. Thus, they are subject to ordinances and the code as specified, unless it is a requirement of the insurance company which insures your home and/or buildings. But we hasten to point out that it is good sense to have your home properly wired and thus protected from electrical fires and damage, no matter where it is located.

In rural areas and outside city limits, you may not be subject to a city inspector's license to wire your home, and you may not have to hire a master electrician. But if you do it yourself, be sure you understand what you are doing, how to do it and what parts are acceptable. In this book we will give you guidelines and examples which, if carefully followed, will help you reach this objective. If you are not sure, get an electrician!

## APPLICATION FOR ELECTRIC SERVICE

You must make application to the electric company, or your contractor must do this, in order to get the electric service connected. Plan far ahead on the date you need the service so that it will be accomplished at the right time. All electric companies request this consideration.

The electric company needs the exact location of premises, including the building street address or lot and block numbers and the name of the subdivision (if appropriate). If you have a new home going up, then use some kind of sign to prominently display the future street name and number.

In the application should be information on the type of service, including the voltage, phase and any special requirements for the load you will impose. The general nature and estimated amount of the residential load is covered. All this information includes the voltage, rating and number of meters and other items to be used. Then the company needs the approximate date that you want the installation ready for service. You will also need the name of the electrical contractor doing the installation. Normally the electrical contractor will take care of all this for you.

Before work is started on a customer's location, the customer or his contractor should secure from the electric company a service outlet and data statement. This will verify what the electric company is going to do and what voltages they will provide. You may prevent some mixups later on if this is done.

In the agreement for service the company will state that it will provide only one type of service (voltage, frequency, phase, etc.). The customer's installation is to be such that all electric service under this agreement can be supplied at one point of delivery and measured by one meter. In the case of apartments and condos, each customer is a resident of an individual unit. Although there are many meters on the building, there is only one which applies to a particular customer.

## WIRING INSPECTION

Before approving any electrical installation, be aware that the first requirement is the wiring conform to the National Electrical Code. It must also conform to the National Electrical Safety Code. Once these are met, then the wiring is checked to see that it conforms to all state and municipal requirements in force at the time the installation is made.

When you are within the city limits, it is probable that certain city ordinances will require that the electric company see a certificate of inspection and acceptance, or a permit, from the proper city authority before they make the service connection. The electric company may refuse to make the connection to any installation it feels or determines is unsafe or a hazard. However, if you have such a situation and have applied for electrical service, the electric company will probably so notify you and may specify how the installation is hazardous or unsafe. In any event the company will delay connection to their lines until they believe the installation is safe. Be aware, however, that any electric company does not assume any responsibilities concerning the customer's wiring or its safe condition.

## MORE CUSTOMER RESPONSIBILITIES

If you are building, remodeling, repairing or adding to an installation, you have certain responsibilities which must be fullfilled. Figure 3-11 shows one home under construction. It is during this phase of the construction that the electrical wiring, plumbing, air conditioning, telephone lines and door bell lines are all installed.

Now let us examine what the electric company does and does not do in further detail, and find out what our responsibilities are in connection with the installation of electric service and the use of the electricity supplied.

First, the electric company does no electrical wiring on the customer's premises other than the installation of its service drop and metering equipment. Second, the customer is solely responsible for any accidents, fires or failures resulting from the condition

Fig. 3-11. A new home is under construction.

and use of his wiring installation or equipment. Since you have this responsibility, and normally you will contract to have the wiring performed, you'd better be sure of the reliability, workmanship and responsibility of your contractor.

Another customer responsibility is to check carefully to see that the phase connections and rotation are correct when first starting motors, and to see that three-phase motors (if used) are not "single phasing." Have your contractor run all motorized devices and see to it that they are running correctly. Normally you will not have three-phase motors in home or residential areas, unless it be an apartment complex, some large condos or a very large home. Checking for "single phasing" is done by your master electrician by measuring the current drawn from each wire of a three-phase connection.

It is your responsibility to plan ahead. Electric service users which may be homes, businesses or manufacturing plants almost, without exception, are increasing the use of their electrical equipment and demanding more service from the lines. For this reason, it is wise to plan for some spare capacity in wire size (for example, use 12/2 instead of 14/2). Plan for the addition of circuits. Be sure that you plan for service entrance facilities so that you will be able to accommodate future requirements which may come from more loading on existing circuits, or the addition of new circuits, and other alterations to existing circuits.

Always remember that you can obtain the latest and most up-to-date information on residential wiring through your electric company. They are anxious to help you, both for their benefit and yours.

To assure maximum safety, *the customer* must provide an adequate and permanent ground conductor attached to the neutral

Fig. 3-12. Switches, raceway and meter box on a larger installation. The meter box is to the right.

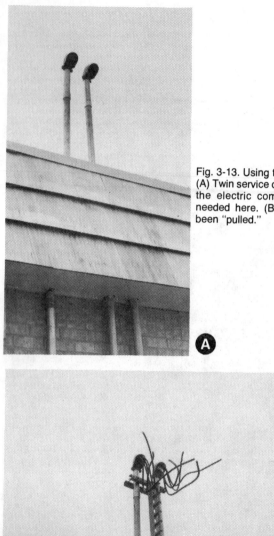

Fig. 3-13. Using twin service drops. (A) Twin service drops to connect to the electric company's lines are needed here. (B) The wires have been "pulled."

terminal of the main line switch or breaker box where a main line switch is not required. Figure 3-12 shows a large switch box being wired and beside it, with just the edge showing, is the metering box fo this installation. Below is a raceway to contain the large wires going to other meters and circuit breaker boxes. On the wires to the right you can see the white color designations of the grounding or neutral leads. Because of the size of these wires, all are stranded. Otherwise you couldn't bend them easily and wire them properly. Clamp-type connectors and pressure fittings are used with this size wire.

In Fig. 3-13 you see the twin service outlets which will be used to hold the wiring that will be connected to the electric company's service drops. See also Fig. 3-3.

A switching panel and conduit risers which go through the roof edge are shown in Fig. 3-14. The large box will contain the

Fig. 3-14. Switch boxes, raceway and meter panel box in a neat installation.

metering equipment. Notice the raceway underneath the two switch boxes. Each piece of equipment is closed to the weather and very solidly mounted to the side of the building. Sometimes, having the switch boxes on the outside of the building helps to cut off the electrical power in case of fire inside.

For all service entrance conductors the grounded neutral conductor shall be electrically continuous from the service outlet through the meter loop. The grounded conductor shall be positively identified either by the use of white insulation, white paint at the terminals, or by some other suitable method. The size and installation of the grounding conductor was shown in Fig. 3-3.

## CHANGES TO EXISTING WIRING

When planning changes, additions or alterations to wiring, the customer shall notify the electric company. Most building alterations or rewiring work will necessitate some change in the electrical input facilities. Of course, any additional wiring must conform to the National Electrical code. In localities having electrical ordinances, there must be approval by the city inspector before the electric company will reconnect its service input. Also, in the case of changes to a building or its facilities, the electric company may find that it is better to change the point of delivery from that used in the past. Consulation with the electric company representative will help to clarify this situation.

If there is a change in the point of delivery, the electric company may require the customer to pay for that change. This will be true if the customer requests a change in the facilities. The payment must cover location of service drop, location of metering equipment, etc.

You can always ask the electric company in your area to help you consider the size and design of the electrical equipment which you may plan to operate with electric power. They can also provide guidance on homes, etc. It is always worth the time and effort to obtain their assistance. You want to be sure your electrical equipment is the proper size and capacity for what you want it to do.

## UTILIZATION VOLTAGE

This is the voltage at the line terminals of the devices you will operate. It may or may not be exactly the same as the service voltage which is the voltage at the point where the electric systems of the supplier and the user are connected. That is the point of delivery. The utilization voltage *may* vary with the location of the utilization equipment. In practice, the service voltage may differ from the nominal voltage. Long lines which produce voltage reductions due to line resistance may be one reason for the difference in the voltage at a source and at a terminal.

## MOTOR PROTECTION DEVICES

All motors need some kind of electrical protective devices to safeguard them, their wiring, and anything which they operate. You might have damage—if you do not have this protection—due to overloading, short circuits, single phasing or large fluctuations in the supply voltage. Fluctuations in voltage may be caused by additional heavy loads coming on the lines at the time you operate your motors. Lightning may strike and cause a rapid sudden rise in line voltage.

Usually an electric company has its system designed to provide a high speed closing of the protective devices on its high lines following a power interruption from lightning or other causes. In most cases these power interruptions will be of an extremely short duration, probably less than 1 second. However, undervoltage protection should be equipped with time delay devices to permit motors to "ride through" these short interruptions without damage.

Notice then that motors need both overcurrent protective devices and undercurrent protective devices. The overcurrent protective devices should be provided in each phase to afford some motor running protection of the three-phase, three-wire motors from "single phasing." You might have a relay which can detect a single phasing condition or running of your motor and then cut if off. Electrical contractors have these kinds of protective devices. You need only discuss the problems and your desires with them, or listen to their recommendations, to have a safely wired protective system.

## MAGNITUDES OF MOTOR STARTING CURRENTS

You probably know that all electric motors require a starting current which is quite a bit more than their "running" current. In cases where these starting currents are very large, you can experience a sudden, abnormal drop in the line voltage when they begin operation. You may have witnessed this in a dimming of your electric lights when the air conditioning motors, or heating motors, come on and start operating.

It is always important that a customer's motors do not exceed their specified starting characteristics so that they will not draw more current than you have planned. Sometimes, when motors are not functioning properly or are overloaded due to some change in the equipment which they operate, they will draw more than the expected and specified currents. Then they can cause trouble on the supply line for yourself, as well as other people connected on the electric line.

There is a specification for the maximum starting current permitted for various size electric motors. We need to examine this and be aware that a master electrician, with his "current loop" meter, can very quickly and easily check to see if any motor is exceeding its permissible current demand. If you suspect trouble due to light dimming or any other kind of symptom such as excessive noise, vibration, or inability to handle the load easily, then have your motors checked.

## Motors With Horsepower Rating

The motors are rated in horsepower:

—For equipment rated at 120 volts single phase or 240 volts, single phase and 2 horsepower or less, the total locked rotor current is not to exceed 50 amperes. Breakers may be used of this value.

—If the motor is rated from 2 to 8 horsepower, then the total permissible locked rotor current is not to exceed 60 amperes.

—For currents permissible on motors over 8 hp, you need to consult with your electric company.

—For 240 volt three-phase motors of 2 hp or less, the permissable locked rotor current is 50 amperes.

—On three-phase motors of 2 to 25 hp, you may have from 50 to 64 amperes.

## Motorized Equipment Rated in BTUs/Hour

When motorized equipment is rated in BTUs per hour, as for air contitioners, heat pumps, etc:

—For equipment rated at 120 volts single phase or 240 volts single phase the locked rotor currents must not exceed 50 to 60 amperes; that is, 50 amperes for the 120 volt types and 60 amperes for the 240 volt types.

—If the equipment rating is 20,000 BTU/hr or less, then the locked rotor current is specified as 60 amperes.

—If the equipment is rated from 20,000 to 60,000 BTUs/hr, then the rating for the motors is 60 amperes plus 3 amperes per 1,000 BTU/hr in excess of 20,000 BTUs.

—For the 240 volt, three-phase motors, the rating is 50 amperes for 20,000 BTU/hr, and 50 amperes plus 2.5 amperes per 1000 BTU/hr in excess of 20,000 BTU.

—On big units, 50,000 to 300,000 BTU/hr, the locked rotor current permitted is 125 amperes plus 1 ampere per 1000 BTU/hr in excess of 50,000 BTU.

Do you know what the "locked rotor" situation is? When electric power is supplied to the motor, you have a locked rotor condition if you lock the rotor so it can't turn. It is always true that an electric motor when running will generate a back EMF (electromotive force) which is really a voltage. This will *oppose* the driving voltage and make it smaller, thus causing less current to be used by the motor because of its smaller supply demand.

For those who desire to operate a group of motors from a single motor control, consult with the electric company to determine what the permissible starting currents for the group of motors can be. Find out whether you can easily and efficiently operate them all from a single control device.

## SPECIAL EQUIPMENT INSTALLATIONS

No electric company wants to have equipment attached to its lines which will cause fluctuations in the line supply or cause interference and problems to its other customers. The customer may be required to attach such devices as self-regulators, filters, etc., before the electric company will supply electric service. X-ray equipment may require special equipment to service it from both the customer and the electric company. It is always good to check out with the electric company before installing special wiring on X-ray equipment or any other possible problem equipment.

Large heating devices operated from three or four wire services shall have the elements connected so the load is evenly divided among the wires, or phases, when this equipment is equipped with automatic controls that may cause frequent connection and disconnection of the load. Heating devices should be operated at the rated voltage for which they are designed. It is well for the customer, or his contractor, to consult with the electric company before ordering permanent installation equipment, so that any problems can be resolved and the type of service to be supplied from the electric company can be determined.

## CONSTRUCTION SITE SERVICE

We have stated that a temporary electric service is supplied to permit builders and contractors to use electric saws, drills, hammers, etc., in fabrication of homes and apartments. If a builder or contractor plans to connect any heavy duty equipment such as sanders, welders and tile polishers to the convenience outlets in any home or business when he does repairs, alterations or installations, he must notify the electric company that such connections will be made otherwise, he might be liable for damage to the electric company's facilities.

Of course, no one should attach anything to the electric company's poles, wires, towers or any other facilities used by the company to supply electric power. Even having antennas, radio towers, ropes, banners or anything like that near the electric company's facilities can be very dangerous and is generally prohibited. Stay away from the electric light poles, lines and other such structures.

## FILTERS AND GENERATING EQUIPMENT

Usually an electric company does not want any modulated carriers or pulsed carrier systems sending signals over its service conductors. If you use a home type wired communication system to send your voice from one room to another, you need to provide filters to make sure this kind of signal does not go out onto the electric lines. In many cases we do not find enforcement of this particular requirement, but it may be enforced. Use filters.

There are two other important items with respect to what you can and cannot do with electric power. Did you know that it is prohibited to use the electric company's energy to supply and energize fencing and make it electrically charged? Also, did you know that you cannot use your own generating equipment to make electricity for your purposes unless you disconnect the electric company's input to your wiring or equipment? In emergency situations it is possible to install automatic control equipment which will disconnect the electric company's lines and connect your emergency generator in their place. This is legal.

## ELECTRIC COMPANY SERVICE

The service we're talking about is not how the company acts when you file a complaint or get a bill that might be wrong. The service we are going to examine is the supply of electric power at the home's location. We have already indicated that the electricity will be alternating, that it will be 60 cycles (nominally) and it may be single-phase, three-phase or two-phase. It is usually 110-120 volts or 220-240 volts. In some rare cases the requirements may be for higher voltages, but these are really not of great concern to us in this book. We will say that if you need a higher than normal voltage or plan to draw more than normal currents (100 amperes) then you need to consult with the electric ompany.

## Single-Phase Service

The single-phase service which is almost universal in homes, condos, apartments and small shops is used for lighting, appliances and small motors. If you have a motor which is up to 10 horse power, it may be connected to single-phase service but must be connected to the 208-240 volt supply lines. For service to a 208-240 volt, two-wire motor or any other type of load, the requirement is that the load be served (supplied) by a three-wire service. Two wires are the current carrying wires and the third wire is bare aluminum size 6 or copper size 8 or larger. It may be insulated if the insulation is white. This wire must be connected to the neutral or ground wiring part of the breaker box, fuse panel or switch panel connector system.

Electric ranges, water heaters and other heating devices for residential use are usually designated for single-phase operation. Such heating devices, if rated at 2,000 watts or more, shall be connected at 120-208 volts or 120-240 volts, depending upon the facilities for rendering service at any specific location. Three-wire service wiring must be provided as we have previously indicated.

Some types of service which are generally available for single-phase operation include the single-phase 120 volt, two-wire service. Use this if you do not have more than two branch circuits and if any motors to be run are not rated at over 1 hp (16 amperes). Also, use it where any heating device does not exceed 2,000 watts. Note, however, that this kind of wiring is considered to be obsolete except for signs, signal lights and such small loads.

Use single-phase 120-208 volt three-wire service if your load does not exceed 10 KVA (kilo-volt-amperes). A 10 KVA rating for a voltage of 120 volts would be

$$\frac{10,000 \text{ VA}}{120 \text{ V}} = 83.333 \text{ amperes}$$

Note that the letter K stands for thousands, thus, 10 KVA means 10,000 volt-amperes.

Of course, any appliance which is to be operated on 208 volts must be plainly marked and rated at 208 volts. Motors to be operated on this voltage must have a name plate saying that is their rating and supply requirement.

## Two and Three-Phase Services

Two-phase service is special and for high use customers who use over 200 KVA but not more than 1,000 KVA. When used, there is a requirement that the load be balanced between the two phases.

Three-phase service is for lighting, appliances, motors and other loads of insufficient magnitude to be beyond the limits of the single-phase service. Again, this is for high use, commercial type customers who have a demand for over 50 KVA or use motors rated at over 3 hp.

Where three-phase four-wire service is provided, the customer must provide wiring necessary to permit the electric company's installation of a three-phase, four-wire metering facility. The service entrance conductors (installed by the customer) must provide a fourth wire for metering purposes and a ground connection between their neutral or ground wire system and the company's neutral or ground wire connection. This wire must be at least a size 6 copper wire.

Normally the other wires in the system will be of the same size. Any single-phase loads attached to this kind of three-phase system must be so balanced that their loading is about equal on each phase. This called a three-phase "Y" type service. In a three-phase "Delta" four-wire grounded neutral service for customers using over 50 KVA, the maximum demand must not be 250 KVA single-phase or 500 KVA for the three-phase system. This means a total of 750 KVA is the limit for both type connections at the same time. It is also important in these high demand service installations that the secondary conductors which attach to the company's transformer and run to the customer's equipment not exceed 50 feet. In these installations, it is a requirement that the conductor serving the power load be identified with an orange covering.

For higher voltages and current demands, the use of a four-wire grounded neutral service will be almost mandatory. Since we get into commercial and industrial wiring with this extended range of demand, we won't elaborate further in this book. Consult your electric company engineers if you need more information in this area of service.

## POINT OF ATTACHMENT

The power lines must be a specified distance above the ground as follows: 10 feet above any surface readily available to a person, 12 feet above driveways in residences not subject to truck traffic, 15 feet over commerical areas subject to truck traffic, 18 feet over alleys and driveways other than residential property, and 22 feet above public streets and roadways. The point of attachment shall not be more than 25 feet above the ground unless a greater height is necessary for proper clearance (Fig. 3-15).

Of course, the service drop must not come in contact with trees. If some tree branches touch the wires, call the electric company for pruning. They usually do a fine job and it may be free. In some cases you may find that it is your responsibility to have this done, in which case you must pay for it.

## ANCHORING THE SERVICE DROP

As shown in Fig. 3-15 a solid anchorage is required for the service drop from the electric company's lines. If the building is wood and permits the use of a wood screw-hook, and the structure is capable of withstanding a pull of at least 300 pounds, the twisted

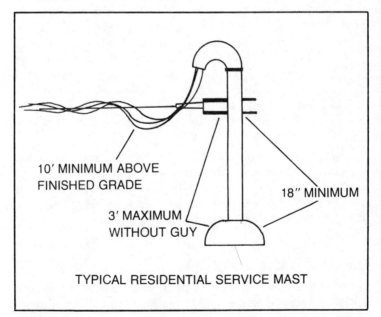

Fig. 3-15. Neutral bare wire supports conductors and is connected to service inlet neutral.

117

Fig. 3-16. A metal rack installation.

service drop support will be furnished and installed by the electric company. Where the wood screw-hook is not practical, then you probably will be required to furnish secondary rack fastened to your home with two ⅝-inch galvanized machine bolts (Fig. 3-16). These must be long enough for the threaded ends of the bolts to extend 2 inches beyond the surface of the wall. Each bolt must withstand a pull of 300 pounds. These bolts must be installed vertically 16 inches apart, with the bottom bolt not less than 10 ½ feet high but not higher than 23 feet. The customer must be sure that his service has the proper clearance. If it does not, check with the electric company for a solution.

If you are planning a new installation which may change the service drop location, be sure your new service outlet location is at least 3 feet away from doors, windows, or porches (Fig. 3-15). If you get a second service drop, it is to be located within 18 inches of the original outlet. Notice in Fig. 3-13 how the two service outlets are side by side.

## SERVICE ENTRANCE CONDUCTORS

Those conductors which must be connected to the electric company's service drop should be installed in rigid metallic conduit. The contractor may use electrical metallic tubing for this purpose. If it is permitted by the local inspector, the contractor may use rigid plastic conduit for the purpose. Each type of conduit must be provided with a weatherproof entrance fitting where they extend from the customer's conduit or raceway. Of course, the connection to the meter socket must be watertight.

The entrance conductors are not to be run through attics, or other enclosed spaces such as in the walls. We have mentioned that at the point where the electric company will connect to these service entrance conductors, you must provide at least 18 inches of length for each conductor where a twisted wire service drop is to be connected. There must be at least 3 feet of wire where an open wire drop is to be connected. You must never have the service entrance conductors in the same conduit or raceway with any other type of wiring. Also, you must never have load wires running in the same conduit with any wires used for metering purposes.

## TRANSFORMER INSTALLATIONS ON CUSTOMERS' PROPERTY

If the electric company feels a transformer installation on your property is necessary, there must be an agreement between the electric company and you. In cases where the homes are being built to be sold, then the electric-company makes an agreement with the builder. These transformers may be above ground on a pad of the type we have previously illustrated. They may, under some cir-

Fig. 3-17. Underground riser of conduit for temporary wiring. Connections are for the builder's electric tools.

Fig. 3-18. Notice the method of connection to electric lines. A clamp is used on neutral bare wires.

cumstances, be located underground in what is called an electric company vault.

## TEMPORARY WIRING FOR BUILDERS

Builders need some kind of connection to an electric supply to operate their tools. The electric company will provide a service drop at a suitable location and also arrange for temporary metering of the electricity used. As with other things of this nature, close cooperation and consultation with the electric company is essential. The safety code of the National Electric Code, and the National Safety Electrical Code must be met. An example of temporary wiring is shown in Figs. 3-17 and 3-18. The plastic underground pipe used will go the electric company's service drop.

## MOBILE HOMES

So many people live in mobile homes that we need to include something regarding the electrical supply to them. Again, here, the customer must have a service entrance with its conductors to connect to the electric company's service drop. This is true especially if this is to be an individual connection to the company's supply lines and the customer is to have his own meter. In many cases, the mobile home owner simply connects into a provided outlet on a pole or stake. There he finds a meter and a cutoff switch.

Many mobile homes use a 12 volt supply for lights and other devices. Sometimes it is a requirement to provide this amount of voltage to the mobile home in a park. Some homes have special transformers installed on them so that the normal 110-120 volt 60 hertz ac supplied from the lines will be reduced to the 12 volts level needed. In these homes there will be a cutoff switch. The switch may be automatic. It will disconnect the electric company supply and connect the mobile home electrical system to its own 12 volt battery and perhaps generator supply in case of power failure or trouble. In no event will the equipment permit connection to both the internal electrical supply and the external supply at the same time.

In some homes it may be possible to convert from the 12 volt supply level to the standard 110-120 volt level. Before doing so, however, you must be very careful and check to see if the wiring is insulated, strong and large enough for this higher voltage. Also, you must check to be sure that appliances connected to the lines will not be damaged by the higher voltage.

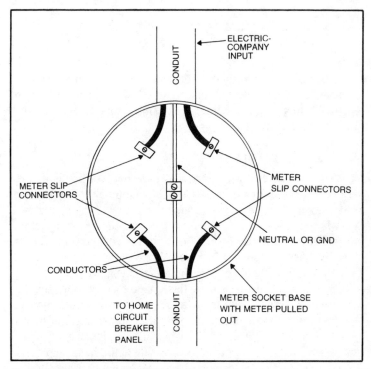

Fig. 3-19. Meter socket wiring diagram for a three-wire system.

### Table 3-1. Meter Socket Base Details.

| RESIDENTIAL SERVICE | | | | | | | | | | |
|---|---|---|---|---|---|---|---|---|---|---|
| TYPE OF SERVICE | | | | METER SOCKET BASE | | | | | | |
| PHASE | WIRE | VOLTAGE | MAXIMUM AMPACITY | DESIG-NATION | CLASS BIN | HUB SIZE | MAXIMUM AMPACITY | APPROXIMATE SIZE W × H × D | | |
| 1 | 3 | 120 240V | 100A | LD-4 | 20-48 | 1¼" | 100A | 8" × | 9" × | 3¼" |
| 1 | 3 | 120/240V | 200A | HD-4 | 20-52 | 2½" | 200A | 12" × | 15" × | 4½" |
| 1 | 3 | 120/240V | 200A | HDU-4** | 20-54 | ** | 200A | 12" × | 15" × | 4½" |
| 1 | 3 | 120 240V | 400A | HC-4 | 20-82 | 3½" | 400A | 13 13/16" × | 26" × | 6" |
| 1 | 3 | 120 240V | 100A | 2 GANG* | 20-56 | - | 100A | 24" × | 11" × | 3¾" |
| 1 | 3 | 120 240V | 100A | 3 GANG* | 20-57 | - | 100A | 32" × | 11" × | 3¾" |
| 1 | 3 | 120/240V | 100A | 4 GANG* | 20-58 | - | 100A | 40" × | 11" × | 3¾" |

| COMMERCIAL SERVICE | | | | | | | | | | |
|---|---|---|---|---|---|---|---|---|---|---|
| TYPE OF SERVICE | | | | METER SOCKET BASE | | | | | | |
| PHASE | WIRE | VOLTAGE | MAXIMUM AMPACITY | DESIG-NATION | CLASS BIN | HUB SIZE | MAXIMUM AMPACITY | APPROXIMATE SIZE W × H × D | | |
| 1 | 3 | 120 240V | 100A | LD-4 | 20-48 | 1¼" | 100A | 8" × | 9" × | 3¼" |
| 1 | 3 | 120 240V | 200A | HD-4 | 20-52 | 2½" | 200A | 12" × | 15" × | 4½" |
| 1 | 3 | 120 240V | 200A | HDU-4** | 20-54 | ** | 200A | 12" × | 15" × | 4½" |
| 1 | 3 | 120 208V | 150A | MD-5 | 20-50 | 2" | 150A | 7½" × | 14" × | 3¾" |
| 1 | 3 | 120/240V | 100A | 2 GANG* | 20-56 | - | 100A | 24" × | 11" × | 3¾" |
| 1 | 3 | 120/240V | 100A | 3 GANG* | 20-57 | - | 100A | 32" × | 11" × | 3¾" |
| 1 | 3 | 120/240V | 100A | 4 GANG* | 20-58 | - | 100A | 40" × | 11" × | 3¾" |
| 3*** | 4 | 240/120V | 150A | MD-7 | 20-51 | 2" | 150A | 7½" × | 14" × | 3¾" |
| 3*** | 4 | 208Y/120V | 150A | MD-7 | 20-51 | 2" | 150A | 7½" × | 14" × | 3¾" |
| 3**** | 4 | 480Y/277V | 150A | MD-7 | 20-51 | 2" | 150A | 7½" × | 14" × | 3¾" |
| 3**** | 3*** | 480V | 150A | MD-5 | 20-50 | 2" | 150A | 7½" × | 14" × | 3¾" |

NOTES:
1. *FOR USE ON MULTIPLE INSTALLATIONS ONLY. MAXIMUM LINE SIDE AMPACITY 200 AMPS. MAXIMUM INDIVIDUAL LOAD CAPACITY 100 AMPS.
2. **MAY BE USED FOR ALL UNDERGROUND 100A AND 200A 1 PHASE SERVICES. HAS 2" KNOCKOUT IN BOTTOM INSTEAD OF HUB IN TOP.
3. ***REQUIRES A BONDING CONDUCTOR.
4. ****3 PHASE COMMERCIAL SERVICE LIMITED TO MAXIMUM AMPACITY OF 150 AMPS IN SOCKET BASE.

If you are buying or thinking of buying a mobile home, it will be important to verify whether you can make this voltage conversion or not. You might, at sometime, want to make this change from the lower voltage to the higher voltage.

## METER SOCKET DETAILS

A meter socket is wired in a simple manner for a home. Figure 3-19 shows the basics of the wiring. Notice how the common or ground conductor goes right through the center of the meter socket, while the conductor wires connect to the slip type connectors into which the meter is pressed. In Table 3-1 we give some details of meter socket bases for residential and commercial services. These bases are obtained from the electric company.

We've reached the end of this chapter concerning the requirements, services and installations of the electric company. Now you know more about where electricity comes from and how it gets to you. We must move on now to consideration of some other things, such as the use of aluminum wiring in the home and other details about wiring.

# Coping With Dangerous Wiring Situations

Dangerous wiring includes those situations which cause problems in electrical circuits and electrically-driven devices. We are not discussing at this time those areas of dangerous wiring which result from incorrect amateur or do-it-yourselfer efforts. Some wiring problems result from nature. Residential and business flooding can cause difficulties with a wiring system.

Lightning and age effects on electrical wiring will also be considered. These involve some techniques which are used to get electricity from "here to there." Sheathed cable should be commonly used, not extension lines. Rigid conduit or special metal surface conduit might be used to give maximum safety. Or you might be able to use flexible armor cable (BX) or conduit instead of the rigid types in residential, condos or apartments.

There are many variations in the types and techniques wiring or "getting the wires from here to there." Generally, however, the use of outlets, boxes, circuit breakers and the manner of running the wires from the service panel to the appliance outlet or where-ever will be about the same. Sometimes there are problems which can seem very difficult when trying to get an outlet in where it is almost impossible to get the wires run properly. It can be very difficult trying to replace bad wiring with new wiring, especially if you want to follow the same channels or locations of the old or bad wiring. Often you just cannot do it economically. It might mean tearing out wall boards, and ceilings and redoing the whole thing. That might be far too expensive. It might be easier to disconnect the bad wiring and leave it in place. Then run new wiring in new locations. First, let's examine aluminum wiring.

## ALUMINUM WIRING

Avoid aluminum wire if possible. In modern wiring it is now proper to use only copper wire or at least copper-clad aluminum wire. We don't recommend the copper-clad wiring. Be safe and be sure. Use pure copper wiring and you cannot go wrong.

Aluminum wire is considered so dangerous that the U.S. Product Safety Commission in Washington, D.C., has published a booklet which says that aluminum wiring presents a serious fire hazard. The commission points out, however, that its viewpoint is open to contest by defendants in a lawsuit. These defendants may be builders who used this kind of wiring in homes, especially during the period from 1964 to about 1973.

What makes this type of wiring so dangerous? It just doesn't make a good electrical connection to switches, outlets, etc., which are made of copper. There is a difference in the heating coefficients of the two metals. If there is enough electrical current passing through a copper-aluminum joint so that heating occurs, the aluminum expands faster and more than the copper. You have a loose electrical connection between the two parts. This, in turn, causes more heat which might cause a fire. It is also true that aluminum wire tends to oxidize more than copper wire. A coating forms on its surface which makes any connection to it, that is exposed to air a "high resistance" type connection and one which can generate heat. On circuits of 15 to 20 amperes, the overheating of the wires and joints may be a problem.

### Checking Your Home For Aluminum Wiring

You should not plan to use aluminum wire in your home, office shop or apartment today. If you have purchased an older home, you might find that it was wired with aluminum wiring. To find out, you can open a wall outlet by carefully removing the surface plate. Then unscrew the two holding screws which fasten the receptacle in the box. Be sure the electrical power is turned off at the circuit breaker box before doing this.

Gently pull the receptacle out of the box until you can see the wires which are fastened to the sides of the unit. If those wires are both silver in color, then you probably have aluminum wiring in your home. Copper wire will be a dull brass or "copper" color. It is easy to see the difference. See if there is any length of plastic sheathed cable in the box. You can examine it in the circuit breaker panel by removing the cover plate over the breakers. You might be able to read on the cable itself the word aluminum or the letters AL.

If you believe your place has aluminum wiring as far as *you* can determine, call a master electrician to verify your suspicions. You will probably need his services anyway, unless you are a good do-it-yourselfer. There is something that can be done if you find that you have aluminum wiring, which will make things about as safe as possible. Before we examine what can be done, let's look at more signs of trouble.

## Trouble Signs

- Cover plates on switches or wall outlets which are warm or hot to the touch.
- Smoke coming from an outlet or switch.
- Any signs of sparks at switches or outlets.
- A smell of burning plastic or other strange odor.
- Lights which flicker periodically without cause. If you have heating and cooling motors, these may cause some dimming when they start.
- The finiding of a switch or outlet which doesn't work may be the sign of an aluminum wiring problem.

You may find an outlet which has nothing connected to it smoking or getting hot. The wiring is through this outlet and on to another outlet which does have something plugged into it. The first outlet's wiring is faulty in the "pass-through" connections, thus, heat is produced. Due to the way aluminum wiring connections change with time in some cases you may find that you have trouble even though things were fine for years.

## Combination Wiring System

There is always the possibility that you purchased your home and it had aluminum wiring. Then later you or someone might have changed some of the wiring to copper, or added rooms or circuits which were in copper wire. Now you have a combination-wired house. If proper connections are not made between these two types of wires, you can be in for trouble.

## How To Avoid Difficulties With Aluminum Wiring

The first thing that a qualified electrician or do-it-yourselfer should do is replace all wall receptacles and switches with a type marked CO/ALR. See Fig. 4-1. Second, it is necessary to tighten all screws on any kind of aluminum wiring to a specific tightness (torque). Your electrician should have proper tools to check this tension.

We bought a home which we later found to have aluminum wiring. Being a two-story home, it was not easy to rewire with copper. Our master electrician replaced all wall sockets and switches with proper CO/ALR units and tightened them to proper torques. He also made certain that every screw covering a wire had the wire underneath it in the proper direction and of the right length for the best possible connection. He was careful not to nick or kink the wires. Aluminum is softer than copper and bruises easily. It can cause trouble then because it won't handle the proper currents that it should. Always avoid nicking, scraping or otherwise damaging any electrical wiring.

Our electrician also checked our ovens, heating-cooling units and everything else electrical. He made the proper connections so that there was no danger from the aluminum wiring. In some cases he had to install special "pig tails," which involves splicing a piece of copper wire to the aluminum wire with a special fitting. The copper "pigtail" could then be connected to a copper fixture or terminal without the possibility of joint overheating due to dissimilar metals.

The electrician also checked all light fixtures. They have the spring-loaded "electrical nuts" which maintain a tension on the wires during expansion and contraction. When these joints are coated with a special "grease" to prevent corrosion of the aluminum wire, they work very satisfactorily with small loads like one, two or several light bulbs. He also checked all permanently wired appliances such as the range, garbage disposal, furnace blower and air conditioner. This was a very thorough job. When he finished we were given a written statement that our home was now checked and modified so it would be safe with aluminum wiring. You also should take this method of preventing trouble if you have aluminum wiring and cannot replace it. We have found in our research for this book that it is easy to say, "don't use aluminum wiring, or don't buy a house with aluminum wirng." But that is easier to say than do as we have learned both from research and our own practical experience. Many people have homes wired with aluminum wire and didn't know it before they bought their homes. Also, you don't just sell a home whenever something bad about it is discovered.

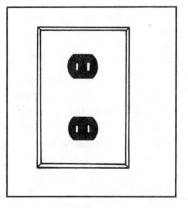

Fig. 4-1. A receptacle for aluminum wiring.

## CHECKING A CIRCUIT

Sometimes it is just a loose or improper connection to a receptacle or switch which may cause the problem. In this case you need to *turn off the power to that receptacle or switch*. Then open up the unit so you can check the wiring and possibly make a new connection to the receptacle or switch which will be correct and extremely tight. When you're done, put everything back together again. If you have smoke or heat flickering from a receptacle or a light on some switch, do the following. First, check the receptacle. Plug a lamp into it and then go to your fuse box or circuit breaker. While you have someone watch the light, remove the fuses one at a time, or turn off the circuit breakers one at a time until you find the one causing that lamp to go out. Then you will have disconnected the power to that circuit. Needless to say, if you have been removing fuses, you will have put them back each time they do not turn the lamp off. With the circuit breakers, you will have reset those which do not turn the lamp off.

Make certain that no one will screw in the fuse or reset the breaker you have tripped. Have someone else watch the fuse panel or circuit breaker box. Put a large sign on the panel which tells people to keep out. You can also lock the panel door with a padlock. Now you are ready to check the receptacle or light switch.

After opening the box and pulling the element out so you can see it, use one hand and turn the element carefully each way so you can see both sides. If there is any evidence of corrosion or a wire which barely goes under the screw head, or if a wire comes loose during this movement of the element, then you have a job to do. You must replace the wiring to the switch or receptacle. We would recommend that you get a new switch or a new receptacle marked

CO/ALR even if the old one seems to work. Remember the new adage, "Save a penny and lose the home."

## CONNECTING TO THE SCREW TERMINAL

Take one wire loose at a time, using an insulated screwdriver and pliers which have insulated handles even if the electricity if off. Try not to touch the other wires on the other side of the unit or any metal with your other hand or any part of your body. Once you have one wire loose, clean it. Cut it back if it is nicked so that you have a nice clean end. Place this under the screw on the new unit, wrapping it in the direction the screw turns. Make the loop in this end of the wire long enough so that it touches itself after going around the screw, but does not overlap itself. Look at Fig. 4-2.

You must do this for each wire connected to each side of the unit. Sometimes there are two wires on each side, and frequently there is just one wire on each side. Now tighten down the screw as tight as possible without breaking the unit or causing the wire to come out from under the screw head. You probably will have to cut off the old stripped end of the wire and restrip and clean a new end to make the connection. Be careful not to nick the wire when you do this.

After you have accomplished the change of unit and have all connections properly made, carefully put the unit back in its box. Move the wires behind the unit into the back of the box. Gently but firmly press the unit into place so it can again be fastened in the box by its holding screws. We have found such a jumble of wires in some boxes that getting the switches or receptacles in place is a problem. Be careful if there are more wires than those connected to the unit itself. These circuits may still be electrified. If you open your box and find lots of wires, go out and pull the master switch, the master fuse or trip the master circuit breaker and cut off all power to the residence. Work your unit back into place without damaging or causing troubles to other wires if they are there.

Sometimes we have found that the wires to the unit may be so short that it's hard to get the unit out of the box far enough to work on it. It is almost impossible to cut a new tip on the wire and fasten it to the new unit. In this case you may have to splice a section of wire—a pigtail—to the installed wiring to make it long enough for this electrical connection to the receptacle or switch. You must be careful doing this if you have aluminum wiring.

You might just take a new section of aluminum wire and splice it into the existing wire. Twist it at the ends and then tape it, or use

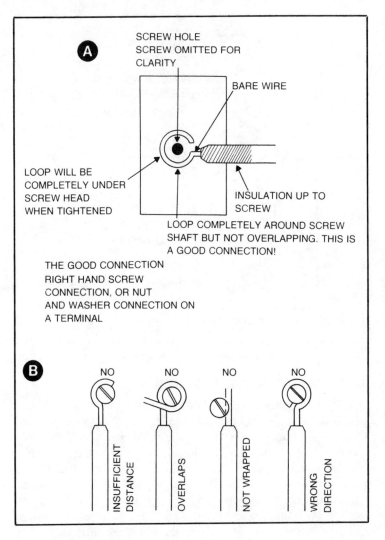

Fig. 4-2. Connecting wires to receptacles switches and screw type terminals. (A) Good connection. (B) Bad connections.

an "electrical nut" which has an insulated cap. The cap is better because of its spring tension connector inside. It is best to use an appropriate wire connecting sleeve which is designed for aluminum (or copper if you are working with copper wire). Put it in place with appropriate tools to make a thorough and safe job. Check with your electrician or electrical parts stores to find out about these units.

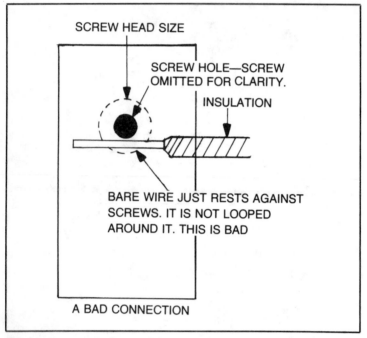

SCREW HEAD SIZE

SCREW HOLE—SCREW
OMITTED FOR CLARITY.

INSULATION

BARE WIRE JUST RESTS AGAINST
SCREWS. IT IS NOT LOOPED
AROUND IT. THIS IS BAD

A BAD CONNECTION

Fig. 4-3. A very bad way to make a screw connection.

## TWIST AND ELECTRICAL NUT CONNECTION

With copper wire, we would consider it safe to use a "twist and electrical nut" connection to be satisfactory. Many examples of these connections will be shown later in this chapter. If you have aluminum wiring, be careful. In Fig. 4-3 we show one example of a bad connection to a screw type terminal. We have also found some connections which were not screwed down at all. The electricians who did the job were sloppy and in a hurry, or their helpers did the job and everything was not checked. Any receptacle or switch which is designed so you just strip the end of the wire and push it into a hole in the back or side to make a connection should be avoided like the plague.

Examine Fig. 4-4 to see how the "electrical nuts" terminals look after the wires have been twisted together. These plastic devices with a metal spring inside have been "screwed" onto the exposed ends of the wires. When properly done, therre are no bare ends or lengths of wires visible on the black (hot) wire or the white (common) wire. The only bare wire exposed will be that "third" wire in such cable(s), which is the grounding connection.

Notice also that there are three cables, each a size 14/2 with ground wire, coming into the box. All white wires have been twisted together. Two of the black wires are left for connection to the switch or receptacle. One black wire is connected through the box, using the electrical nut to prevent any undesirable connection to anything else in the box. There may be two switches placed in

Fig. 4-4. Electrical nut insulators and connectors cover the ends of the twisted wires.

this box. This is probably the case because the box was located about 4 feet high on the stud. Boxes which are within 2 feet of the floor are for receptacles. Higher ones are for switches, except in very rare cases. This kind of box connection might be for an outside front light and an entry light, so two switches are needed.

In Fig. 4-5 three of the wires from a single cable are shown: the black, white and the bare ground wire. Notice the cable designation (14/2G) and the way the cables go into the box. Sometimes fittings are used and sometimes they are not. We were informed by the electrical contractor that *if* the box is plastic, there is no need for the fitting connection. If a metal box is used, then the cable termination and box connection fitting is mandatory. Also note how the box is fastened to the stud. A special channel at the top permits a nail to hold it in place. A second nail, similarly located, is on the bottom of the box. Of course, three of the wires (from the second cable) have been tucked back into the box so they cannot be seen in Fig. 4-5.

## SMOKE DETECTORS AND FIRE EXTINGUISHERS

The U.S. Consumer Products Safety Commission recommends that anyone having a home with aluminum wiring install smoke detectors and have on hand a fire extinguisher designed for use on electrical fires. If a fire extinguisher can put out an electrical fire, its label will say so. Check to make sure you have the proper type. Regarding the smoke detector, get one with the two types of detection elements, photoelectric and ionization, within its case. Check the unit periodically say every three months, to be sure it is working properly. Replace batteries and even the unit itself if it doesn't work when tested.

Do not install a smoke detector and then forget it is there. Smoke detectors are a cheap but very effective insurance if they are maintained. Use as many as are necessary if your home has aluminum or copper wiring.

## RECOMMENDATIONS FOR ALUMINUM WIRING CONNECTIONS

Never make a "hot" connection. Turn the power off. Don't use unmarked devices (switches, receptacles, etc.) when replacing or making connections to aluminum wiring. Never use the old style devices which may be marked AL/CU, use CO/ALR. Don't use the "push in wire" connection devices or any other "quickwiring" or "back wiring" techniques with aluminum wiring. Connect aluminum wire only to wrap-around wire binding terminal screws

Fig. 4-5. Shown are wires in a plastic box and the mounting of the box.

on devices which are marked CO/ALR (Fig. 4-2). If you have to replace a receptacle or switch and you have aluminum wiring in your home, then do the following:

■ Remove the wire insulation so that you obtain a fresh surface of the wire itself. Do this without nicking the wire.

■ Wrap the wire the proper distance around the terminal

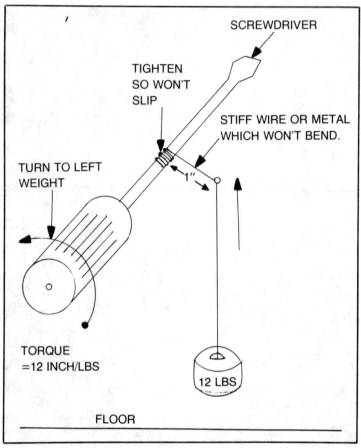

TIGHTEN
SO WON'T
SLIP

SCREWDRIVER

STIFF WIRE OR METAL
WHICH WON'T BEND.

TURN TO LEFT
WEIGHT

1″

TORQUE
=12 INCH/LBS

12 LBS

FLOOR

Fig. 4-6. The torque here equals 12 inch-pounds.

screw. Be sure it goes in the proper direction.

■ Use the proper torque to tighten the terminal screw. Tighten the screw as tight as possible without damaging the unit or cutting deep into the wire.

When you replace the unit in the box, carefully move the wiring so the unit will go into place without pulling on the wires or loosening any connections.

## TORQUE READING

Some screwdrivers are so designed that they give a torque reading. The proper screw tension for aluminum wire is 12 inch-pounds (Fig. 4-6). One way you can check this, if you do not have a torque type screwdriver, is to fasten something to the blade of the

134

screwdriver so that it projects 1 inch from the shaft. Now tie a string or cord to the end of this projection. On the other end fasten a weight—12 pounds would be just right. To get a "feel" for this tension, use a 10 pound sack of sugar or something that weighs between 10 and 12 pounds fastened to the other end of the cord or string. Now twist the screwdriver handle until the weight is just off the ground or floor. That is the tension needed.

If you do not have a torque type screwdriver and you don't get a feel for the tension as suggested, tighten the screw. Tighten it until wire is firmly in contact with the bottom of the screw head and the contact plate on the CO/ALR unit to which the wire is being connected. Now, tighten the screw an additional half turn to flatten the wire between the screw head and the contact plate. See Fig. 4-7.

If you have a choice, always use copper wire. Sometimes you don't have that choice because the home you bought has aluminum wiring, and you can't change it for a variety of reasons. We have given you the best available guidelines for handling aluminum.

## ELECTRICAL FIRES

Electrical fires usually start out as small fires. The first precaution is to have one or several fire extinguishers in the home.

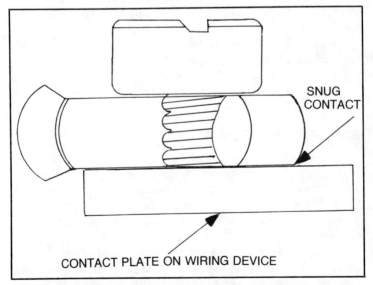

Fig. 4-7. How to tighten aluminum wire to the proper torque. Tighten the screw until the wires are snug as shown. Then turn the screw ½-turn more.

If the garage is detached, it might be well to have a small one there. Second, you should make certain that the extinguishers will put out the kinds of fires which may be encountered. You need at least one for *electrical fires*, and you might have another which is for fires caused by burning paper, wood, plastics, cloth or rubber. A good type is a pressurized water extinguisher. You must tackle a fire immediately when it starts if at all possible.

If the fire reaches any size at all, call the fire department. Don't wait until it is a *big* fire. Call in time for them to save your home.

Some small fires in the home may be handled in the following manner. If you detect or smell smoke coming from an appliance or any electric motor, pull the plug or turn it off. If any flames appear after this is done, use water to extinguish them if you do not have an extinguisher.

A fire may develop in an oven where food may catch on fire. Close the oven and turn off the oven heat. Do not open the door and fan the flames. That might cause the fire to spread outside, and then you've got real trouble.

If you detect smoke coming from the television turn it off or pull the plug. Stay away from the television in such cases as the picture tube may blow up.

Sometimes a fire develops when cooking on the top of the stove. Cover the frying pan or whatever with a lid, if possible. If the fire is "down below," then call the fire department immediately.

Finally, if your clothes catch fire, don't run. Lie down and roll over and over. Wrap up in a carpet, blanket, coat or anything which can smother the flames. Try to get you clothes off if they keep burning. Don't pull them over your head so your hair catches fire or your eyes become injured.

## MORE FACTS ON SMOKE DETECTORS

We will expand a little on what we said about smoke detectors earlier. Most smoke detectors have either of two types of units to detect smoke, a photo-electric cell arrangement or an ionization chamber type detector. We suggested that the best type of smoke detector is one which has both these units incorporated into its mechanism.

Some smoke detectors run on batteries and others are designed to run from the electric current in the home. These latter ones are permanently connected into the house wiring. As you can guess, the house current operated detector will be inoperative if

**Table 4-1. Information on Types of Smoke Detectors.**

| Manufacturer | Operation | Detector | Cost |
|---|---|---|---|
| Sunbeam Appliance | Battery | Ionization | $30.00 |
| First Alert | Battery | Photo-cell | $40.00 |
| First Alert | Battery (w/light) | Photo-cell & Ionization | $50.00 |
| Pry-A-Larm | Battery | Combination | $25.00 |
| Westclox | Battery | Ionization | $27.00 |
| Honeywell | House current & Battery | Ionization | $ ? |
| Norelco | House Current | Ionization & Remote alarm | $250.000 |
| Fyrnetics (Elgin, Ill) | Battery (10 Year) | —————— | —————— |

the house electrical supply is interrupted for any reason. A battery operated unit will operate under these circumstances if you replace that battery as often as is necessary. Some units have test buttoms. Others require that you create some smoke from a cigarette, a burning rag or something to test them. Normally you will not have a volt-ohm meter which is used to test the battery in or out of the circuit. With batteries it is a wise precaution to change them every 9 to 12 months just to be sure.

We have indicated that a smoke detector should be placed in the hallways, on each floor level, near doorways where several bedrooms are located, in basements, kitchens and even in the attic. Some people object to placement of a smoke detector in the kitchen. Many have experienced an alarm due to smoke from the cooking when no fire was present at all. You have to decide about this location. Table 4-1 has information about some brands of fire detectors.

Most fires occur between 10 pm and 6 am according to some studies on the subject. Detectors in bedrooms are essential. Some people don't wake up easily and need to have that loud alarm blast from the detection unit to get up in time. Do not regard smoke detectors as luxuries. Consider them to be necessities and use them no matter what kind of wiring you have in your residence. If you have aluminum wiring and cannot change it, then a smoke detector is probably a vital necessity.

## CHECKING WIRING IN A WATER-DAMAGED HOME

Since this chapter is concerned with problem avoidance in electrical systems, and electrical system safety, it is not inappropriate to consider the situation which can occur if your home or other buildings with electrical systems that have been flooded or otherwise subject to water damage. Assume that you have left the premises and now return or plan to return. The first step is not to turn on any electric circuit, or touch or handle wires attached to lamps, appliances, etc., without first having the residence completely checked by a master electrician for short circuits. It is entirely possible that some insulated wires may now have developed electrical leakages. If you handle them, especially if the floors are wet or damp, you might be electrocuted. You do not want to handle or snap the master circuit breaker switch or try to run the electricity off at the home's entrance if there has been water. Stay clear of everything until it has been checked.

A master electrician or someone knowledgeable about electricity can help if it is vital to get into the home. This person should follow some very important and basic precautions. First he should wear rubber-soled shoes or wading boots. Then he should have a good pair of rubber gloves. When he has taken these precautions, he should try to open the circuit to the residence by turning off the main switch wherever it is located. As this is done he should stand on a dry board and not touch anything metallic around that electrical connection. A dry stick might be useful in snapping the main switch open or snapping the main circuit breaker to the off position.

Never underestimate the danger around a fuse or circuit breaker panel if water has been inside. Water in the conduits can leave deposits which may conduct enough electricity to cause death. There may be short circuits in the wiring. Fires and sparks may occur if anything is moved in the box.

The electric company workers will normally be on the scene in any disaster area within a short time. These men will probably disconnect the lines feeding any such waterlogged area. Of course, the normal short circuiting will probably trip the electric company's fuse panels or circuit interrupters, but don't rely on this happening. Instances have happened where a fusing of wires across such current protective devices has permitted electricity to flow even though it was not supposed to do so.

## THE AFTERMATH OF FLOOD DISASTER

It probably will be necessary to rewire much of a residence which has been flooded. This will take time and must be done by highly trained and capable personnel. *Romex*, the plastic sheathed cable which we have mentioned, may or may not be of the waterproof or water-resistant variety. If it is not, a watersoaked cable will have to be replaced. Receptacles, switches and appliance connections will all have to be replaced and connected to the new wiring. It isn't a pleasant concept to think about, but it will pay you to have the job done right. In some cases the electric company will have representatives who will make a check of your home's new wiring and give you the okay, or not, to continue with their service. You may have to ask for this check, and it will be worth the effort.

Any underground wiring, yard lighting and swimming pool lighting is normally installed in waterproof type conduit, water-resistant or waterproof cables, etc. If the place has been flooded, you must be sure that these cables were not damaged. Everything electrical must be checked out thoroughly. Sometimes after a flood situation a temporary electrical service might be installed to provide power to sump pumps and the like in order to do a cleanup. Working with electrical devices on damp or wet ground can be very dangerous. Again, rubber boots, gloves are a wise precaution against electrocution.

Sometimes you might have outside wiring conduit or fuse boxes which might be dampened by rain. As a result, troubles might occur inside your residence. A light may go out here or there or an appliance may not run properly. Have your home checked for electrical leaks or short circuits. Use of drip loops, electrical cable, and properly sealed conduit is important. If the electrical lines, cables and equipment were installed correctly, there may be no damage. But don't take chances.

If you have small electric motors which have been under water, it is sometimes possible to dry them out in an oven set at not more than 150 degrees. Leave them at this temperature for at least 6-8 hours. After drying the motors out, they need to be tested for shorts. A qualfied person can use an ohmmeter to see if the electrical resistance seems adequate.

If lamps have been exposed to water, remove and replace their sockets, switches and electrical cords. They will probably be as good as new. Some items like an electric iron, or any type of appliance which has a heating element or a thermostat inside it, will probably be so dangerous due to a change in its control and active elements that it should be thrown away. Ovens, electric stoves, and heaters all fall into this category.

## CHECKLIST

The US Department of Agriculture recommends a series of steps which can be followed to check the electric wiring in a home if there has been flooding.

- Put on rubber boots and wear rubber gloves.
- Open the main switch on each building or at each yard pole.
- Remove all branch circuit fuses; open all branch circuit circuit-breakers.
- Disconnect all plug-in equipment and open the switches at each piece of permanently connected equipment.
- Clean the dirt and debris from the load-center panels and the switch, outlet and junction boxes.
- Allow the entire wiring system to dry out. No time is specified for this, and the time may vary according to the humidity in the area.
- Have the whole system checked by a skilled electrician, preferably an electrical inspector who has equipment for testing insulation resistances.
- Make all changes recommended by the inspector.
- Insert the fuse (close the circuit breaker) in a single branch circuit. Close the main circuit breaker or switch. If the fuse blows or the breaker trips, there is still a fault in the circuit. If the fuse does not blow (or the circuit breaker does not trip), look over the visible wires and each outlet in that circuit to see if there is any smoke or other signs of faults. If everything seems to be normal, open the breaker (remove the fuse to that circuit).
- Repeat the previous step for each of the other circuits.
- After all circuits have been individually checked, insert the fuses in their proper places (close all circuit breakers). Plug in and operate an appliance, known to be in good condition, for each outlet.
- Do not connect appliances and equipment that have been submerged until each one has been properly checked and tested. After proper cleaning and checking, each one should cautiously be operated individually to be sure that it is in good condition. If the appliance doesn't run properly, disconnect it immediately and have it tested again. You may have to throw it away.

## CLEANING ELECTRICAL MOTORS

There are many electric motors in a home or residence which you don't often see. For example, there is the vent motor in the

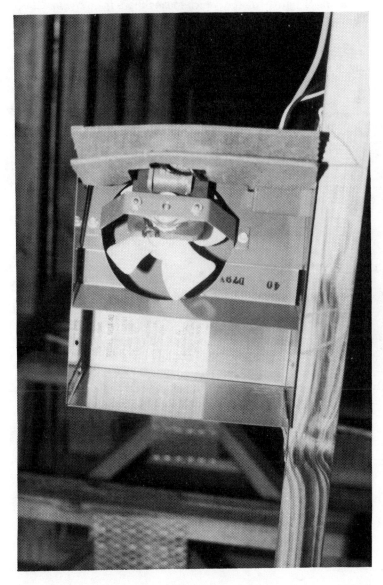

Fig. 4-8. A venting motor unit for a bathroom.

bathroom or kitchen. One such unit is shown in Fig. 4-8 as it is being installed in a new home and before the venting duct has been attached to it.

This kind of motorized unit will normally be concealed in the wall or other inconspicuous place. It may work automatically from a thermostat, or you may have a switch to turn it off and on. It is a *synchronous* type of motor, which means there is no *commutator*. The motor may continue to work properly if the field windings are allowed to dry out if they become wet. The clearance between the armature and the field poles must be carefully checked to remove any mud, dirt or residue.

Another motor is found in the air conditioning system. Of course, there are many types of blower motors used in heating-air conditioning systems according to the size of the house or apartment. One small unit for a one-bedroom apartment is shown in Fig. 4-9.

Notice that the motor may be easily removed for replacement or repair. If the motor is larger and is not sealed, there may be quite a cleanup chore to get it back into shape again. Let us examine one procedure which you might use to clean up an electric motor.

## Taking Apart the Motor

First, turn the power off. Then carefully disconnect the electric wires, taping the bare ends of each with a small piece of white adhesive tape. Mark in ink the connection to which the wire goes. Stick that tape on the wire so it won't come loose. Remove the motor, being careful to note exactly how all parts are fastened in place. Put the screws and bolts carefully aside and try not to mix them up. Label them with a paper tag if necessary.

When you take a motor apart, you have the same situation. If it has a case, then you probably should take an ice-pick or sharp knife and scratch a line across every case joint (Fig. 4-10A). Make sure the scratch lines match when you reassemble everything. If the front and back sections seem the same size and shape, then scratch a "front" and "back" on them so you can remember which is which when reassembling. Any screws or bolts taken out of the case should be carefully marked. If necessary scratch a number or something on the case near the hole where each goes. It is hard to get a motor back together again if you hurriedly take it apart and then find yourself with lots of pieces, not knowing where each goes.

Fig. 4-9. A small air conditioning unit motor is visible.

## Bearings and Brushes

Remove and wash, thoroughly, all bearings that are not sealed. When you replace them, be sure you oil and/or grease them just as thorougly. Clean out the "wells" supplying the bearings with oil and fill them with fresh oil.

Clean centrifugal switches, slip rings, and commutators of any grit, dirt or grime. If the motor has brushes, examine them to be

143

sure they are clean and fit tightly in their holders. When you remove an armature from a motor which has brushes, the brushes will normally spring out and "dangle" near their holders. Be careful not to break any wire connections to the brushes. Some just fit into holders. Others have a small length of copper or brass wire pig-tailed to them to connect to a screw for a better electrical connection. Also brushes will have a curve at the end which makes contact with the armature. When reassembling the brushes, be sure you do not get this curve crosswise to the armature. If you do, heavy sparking will occur for awhile. The motor may not run properly until it "runs in" a new curve in the brushes. See Fig. 4-10B.

When you go to reinsert the armature or put the cap on the end of the motor you may find some difficulty if there are brushes. The brushes will dangle down. If you hold three, the fourth "gets away" to prevent the armature from slipping into place. Try using a piece of cloth, carefully and gently wedged into the brass or copper holder to keep the brush in place until you get the armature reinserted. Be sure you can pull that cloth out easily and surely once this is done, or the brush won't ride free and be able to make continuous contact with the armature when the motor is running. Eventually the brush will have to be replaced.

When you take out the armature or rotating member, clean it with water from a hose under low pressure or put it in a pail of water and scrub it down gently. Do not scratch off any insulation or scratch up the commutator. Some armatures seem to be just a solid mass as their wires are embedded into the iron. These clean easily. Do not use high pressure water or even high pressure air.

## Cleaning and Drying Coils

If the motor has stationary coils in the main section of the body, these can be cleaned in the same manner. Once the motor has been water-cleaned, clean it with kerosene. We recommended that gasoline not be used,   but sometimes it   as kerosene is often hard to obtain. In either case, do not smoke near the fluid. Be sure the coils and parts are thoroughly "air-dried" before drying in an oven.

The motor coils, either *rotor* or *stator*, should be dried by heating to a temperature of about 150 degrees Fahrenheit for 10 to 15 hours. Then the coils should be painted with a light insulating varnish. After painting the first coat, then "bake" it for 4 or 5 hours at a temperature of 200 to 250 degrees Fahrenheit. Apply a second coating and bake it for 3 to 4 hours at that same temperature. You do not paint the commutator or the bright brass or copper-wedged

144

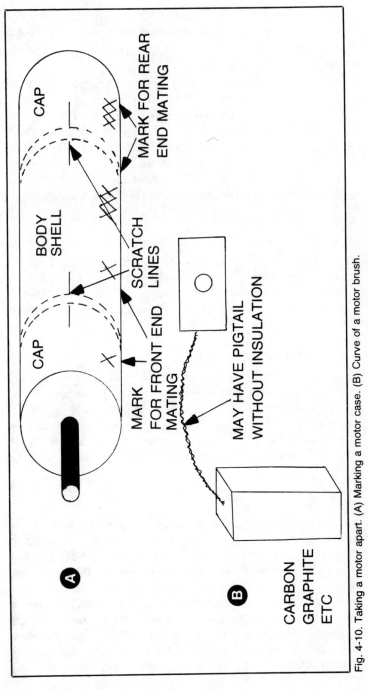

Fig. 4-10. Taking a motor apart. (A) Marking a motor case. (B) Curve of a motor brush.

145

section at the end of the armature. Nor do you paint any brass or copper rings which may be at the end of the armature. You paint only the coils of wire.

## Removing Dirt From Bearings

Before reassembling the motor or generator, check the starting contacts for corrosion and lubricate all moving parts lightly. Replace any oil wicks and renew oil in reservoirs. The bearings should be thoroughly reconditioned. If they are scaled and have leaked, permitting grime and other foreign matter to enter into them, they should be soaked in kerosene, gasoline or oil. Remove the dirt and grime. Compressed air under some pressure can be used to get the dirt out of the bearings. Don't use water to clean bearings.

## Capacitor Type Motor

When you try to run a capacitor type motor, the capacitor may smoke. It just isn't worth the effort to try to clean up a capacitor, so replace it. If the motor case seems slightly corroded or rusted, a good first step is always to replace the capacitor, even if it looks good.

## Running the Reassembled Motor

Once the motor is reassembled and everything seems proper, run it without load at first and then with load. If you have or can borrow a current meter which simply clamps around one of the motor leads, use this to see if the current is about what the factory specifications call for. Depending upon cost and availability, it might be best just to replace the motor. We leave that choice up to you.

## BAD CORDS

Since we have been discussing dangerous electrical situations with aluminum wiring, bad wiring and events due to uncontrollable causes, let's see how to avoid other electrical hazards. Old, frayed and broken electrical cords are very dangerous. Take a few minutes to examine the cords that you have around the house. If they have cracks in them, the insulation may be inadequate. If you repair such cords and lines by wrapping them with tape, often the tape becomes loose and no longer serves its protective and insulative task. Often, right at the juncture of a device and its cord, we have noted insulation breaks so bad that the wires can be seen.

What happens with these old, worn electrical cords? If you are lucky, when the cords finally wear through enough so that the wires can touch, there will be a flash, some sparking, and the smell of sulphur and perhaps burned rubber as the ends of the wire fuse together. A fuse will blow or a circuit breaker will trip. You can make a temporary repair. The ends of the wires are clipped or cut off each segment of the cord. Then they are reconnected, through twisting and taping, and the fuse is replaced or the circuit breaker reset. Later in the book we will explain how to replace a bad electrical cord.

## ANTENNA INSTALLATION

Another dangerous situation is antenna mounting for either a television or CB radio. Place your antenna so that it cannot touch a power line, a service drop, or any kind of electrical wiring. Make certain that the antenna is very securely fastened so it cannot fall in such a way that it will make an electrical connection.

We have witnessed many antenna installations. In many cases the owner uses some kind of pole, aluminum or other metal, which is fastened at some point on the home or nearby structure. Then the antenna is put at the top. Frequently the antenna has no guy wires to the top of the support pole. Always use at least three guy wires on any antenna. Guy wires can keep the support pole from bending or breaking, and they do not interfere with the electrical reception when properly installed.

## LIGHTNING AND LIGHTNING RODS

If lightning strikes an electric line, the chances are high that a circuit breaker installed by the electric company on its lines will open and prevent damage to the people it services. Sometimes, however, and electrical *transient*—a high voltage which exists for a very short period of time—might get by that circuit breaker. This transient will come into your electrical system.

If a transient does come into your home's electrical wiring, it probably will not trip your circuit breakers either. It can get into any devices which you have connected to the electric lines. Turn off all appliances which might be damaged by this high-voltage short-duration transient. You especially want to disconnect any computing equipment as transients are deadly to these devices.

What about lightning rods? Lightning rods are simply metal spikes which protrude above a building at each corner. They are connected by large electrical wiring such as an electrical cable size 8/2 with both wires connected together. Then the other end of this wire goes to a metal spike at least an inch in diameter which is placed about 6 feet into the ground. In the city or a crowded residential area you may have lots of lightning, but the usual situation is that the lightning does not strike buildings. Structures in rural areas are more often hit by lightning.

# Circuits and
# Electrical System Layout

We have discussed a series circuit, used for switching, and a parallel circuit, used to connect wall receptacles. We have considered just what a circuit is in reality. The circuit is a closed loop so that the electricity can come from the source, go through the devices and get back to the source. We have hinted but not discussed the fact that there are *series-parallel* circuits. These are combinations of the two basic types. Now we will expand on these ideas and add something more on switching. The routing of electricity is governed by how we switch its conductors around. It goes the way the tracks are connected. Electricity does not follow a broken track line. So we need to consider the switching and devices used to change various types of electrical connections.

We shall be discussing some techniques used by professionals in wiring and installations, and we can learn from the way they do things. Later in the book we will examine modern home wiring and methods of handling electrical currents. We need to have a good understanding of the concepts in this chapter to get on a common ground as we proceed forward.

### SERIES, PARALLEL AND SERIES-PARALLEL CIRCUITS

It may seem like a review, but we will take some space to examine the series, parallel and series-parallel circuits. Consider Fig. 5-1 where these three important circuits are diagrammed. In Fig. 5-1A we see the simple switch circuit, connected to a light to

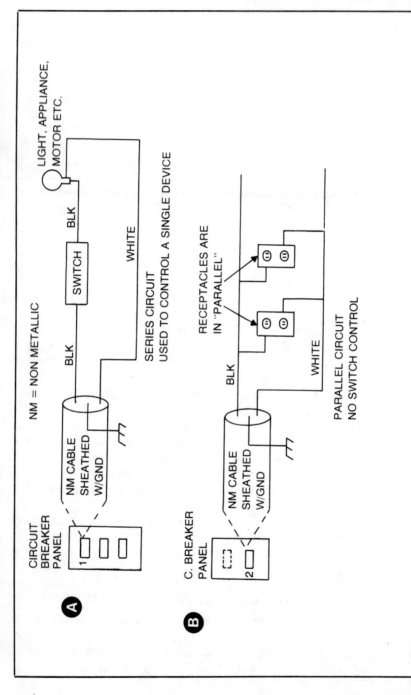

NM = NON METALLIC

LIGHT, APPLIANCE, MOTOR ETC.

BLK

SWITCH

WHITE

BLK

SERIES CIRCUIT
USED TO CONTROL A SINGLE DEVICE

CIRCUIT BREAKER PANEL

NM CABLE SHEATHED W/GND

1

**A**

RECEPTACLES ARE IN "PARALLEL"

BLK

WHITE

PARALLEL CIRCUIT
NO SWITCH CONTROL

C. BREAKER PANEL

NM CABLE SHEATHED W/GND

2

**B**

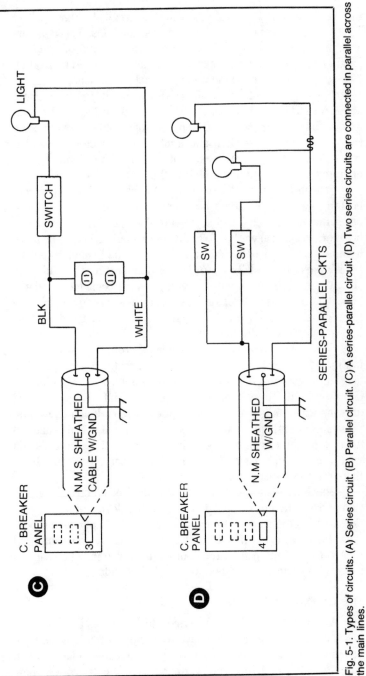

Fig. 5-1. Types of circuits. (A) Series circuit. (B) Parallel circuit. (C) A series-parallel circuit. (D) Two series circuits are connected in parallel across the main lines.

151

turn it off and on. This is always a series circuit. In Fig. 5-1B we see two wall receptacles connected across the lines. This is *always* a parallel circuit. In Fig. 5-1C we see one version of a series-parallel circuit. The wall receptacle sockets are wired in parallel. The light is connected to the wall receptacle terminals through the switch, so this part of the circuit is a series circuit. In Fig. 5-1D we see how two light switches are connected to the cable to the circuit breaker box, and these control separate lights independently. The two series circuits (switch and light) are connected in parallel across the main (cable) lines. Notice that either light may be operated by itself without affecting the other.

Also in Fig. 5-1 we have indicated that each circuit is connected to a separate circuit breaker in the panel. We have numbered these 1, 2, 3 and 4 to give you some "feel" for the use of circuit breakers in general wiring. Actually there would probably be more lights, receptacles or combinations of these on a single breaker. Each breaker is normally set for 15 to 20 amperes for these kinds of circuits. But this simplified case gives you the basic idea, and it is important that you understand these basic ideas to begin with. Now let us consider the diagrams using the generally accepted symbols for wiring (Figs. 5-2A and 5-2B).

While Fig. 5-2 doesn't seem much different than what we had in Fig. 5-1, we include it to indicate to you that you should study a diagram regardless of symbols, and then determine what kind of circuit is used. There should always be a legend on any electrical drawing which tells you what any symbol means. If you know this, you can understand what kind of circuit is used and what the devices on the circuit are like. Knowing what they are tells you what they do, in most cases. Then, if necessary, you can redraw the circuit in a manner you understand.

Check it for short circuits, those undesired electrical connections across the lines which do not have any device in series with them. Check the diagram for continuity, which simply means looking at the loops to see if electricity can get to where you want it and then get back out again to the source. Then you may be ready to install the wiring and connect it with full confidence.

## THE THREE-WAY SWITCH

A switch is a mechanical device which touches and thus connects metal parts inside its body. A simple switch like we have examined in Figs. 5-1 and 5-2 have simply a single "blade" which snaps into a metal receiver section so that electricity can flow

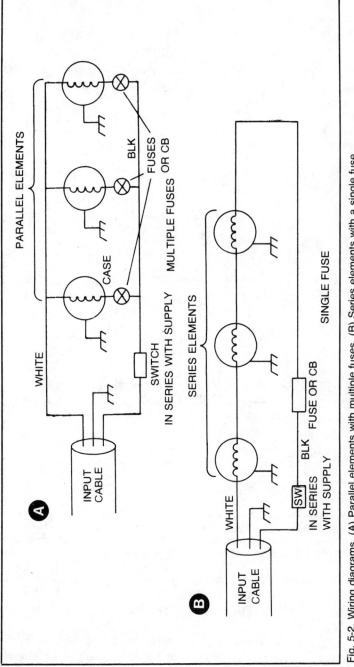

Fig. 5-2. Wiring diagrams. (A) Parallel elements with multiple fuses. (B) Series elements with a single fuse.

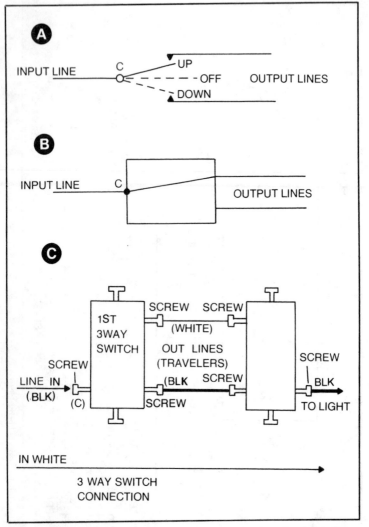

Fig. 5-3. Three-way switch representations. (A) A single pole, double throw switch with the neutral position off. (B) Block symbol of a single three-way switch. (C) This three-way switch has no off position.

uninterrupted. Now we consider the three-way switch. When it is in the "up" position, it connects to one wire input; when it is "down," it connects to a second wire input. This is shown in Fig. 5-3.

In Fig. 5-3A you see the electronics symbol. Figure 5-3B shows the electricians' symbol. In Fig. 5-3C you see a pictorial

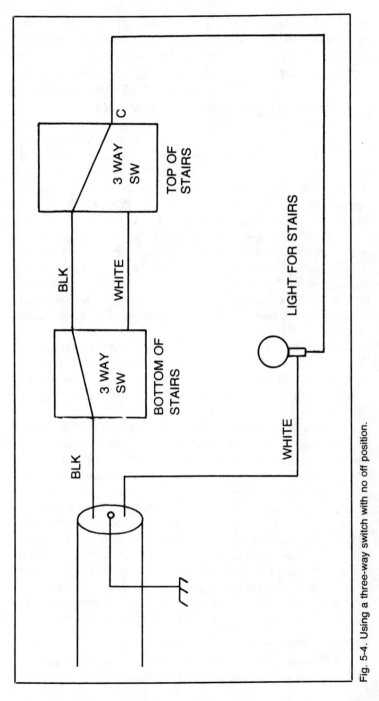

C

3 WAY SW

TOP OF STAIRS

BLK

WHITE

3 WAY SW

BOTTOM OF STAIRS

BLK

LIGHT FOR STAIRS

WHITE

Fig. 5-4. Using a three-way switch with no off position.

which designates the input connection as (C), which is the common connection to that part of the switch which moves to connect to the two fixed metal pieces. These, in turn, are connected through screws to the lines. The importance of this switch is that it has no off position. It connects the input to one of the outputs all the time.

## USING THE SINGLE POLE DOUBLE THROW SWITCH

It is nice to be able to turn on a light from one area and turn that same light off from another area. For example, let's say you can turn a light on at the bottom of the stairs and turn if off at the top of the stairs. Conversely, you can turn the light on at the top of the stairs and off at the bottom. There is a circuit which, when connected to two of these three-way switches, will permit this type of operation. That circuit is shown in Fig. 5-4.

Notice in Fig. 5-4 that the light is on. If either switch is moved, then it breaks the circuit through line 1 and connects to line 2. This will turn the light off. If the second switch is moved to its second position (like the off position on a normal light switch), the light will again come on because the circuit is again completed through line 2. This circuit is common and very important around homes.

This circuit can be used between the garage and the house. Turn the garage light on at the back door; then turn if off at the garage as you take the car out. Turn it on at the garage when you return, and turn if off as you go inside the house. You'll think of other uses of this circuit.

Examining the wiring of this kind of circuit, notice that it takes a sheathed cable with black and white wires and ground wire. You'll use only the black and white wires, although you can connect the ground wire to the ground wires of the other cables for protection. The input (black wire) from the circuit breaker box or the source of power always goes to the common (C) terminal of one switch. The common (C) terminal of the second switch is then connected to the light with a single wire. If possible, it should be black. The other side of the light is connected to the common white wire which comes to the first switch through the input cable. As you see, these aren't difficult to connect together at all. These switches are readily obtainable at electrical departments of hardware stores or electrical supply stores. Just look for three-way switches.

## FOUR-WAY SWITCH USE

A four-way switch has two terminals on each side and connects in between the two three-way switches as shown in Fig. 5-5.

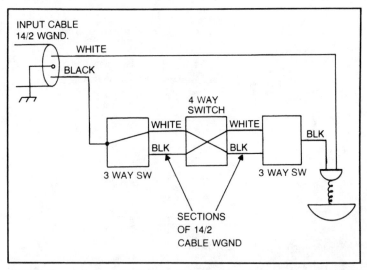

INPUT CABLE
14/2 WGND.

WHITE

BLACK

4 WAY
SWITCH

WHITE WHITE

BLK BLK

3 WAY SW 3 WAY SW

BLK

SECTIONS
OF 14/2
CABLE WGND

Fig. 5-5. Use of a four-way switch to get three independent light control positions in three different places.

You can use two sections of the sheathed NM cable and connect all grounding wires together. The line from the last three-way switch to the light fixture will, of course, be a single wire. Probably the white wire connecting to the other side of the light will be a single wire going back to connect to the input cable common (white) wire. With this kind of circuit, any one of the three switches will be able to turn the light off or on regardless of how the other two are positioned.

## ELECTRICAL SYMBOLS

Before going ahead with more on circuits, let us examine some of the symbols which are in use currently to specify verious types of electrical devices such as lights, fixtures and switches. We show two types of lists because there will be small differences due to which architect or electrical designer has prepared the plans. Examine Figs. 5-6 and 5-7.

We will examine a blueprint later on and will need to have an understanding of electrical symbols, to understand what the blueprint shows us. When examining the "roughing out" methods used in modern electrical installations note the symbols SSS or whatever, written on the 2×4s, which designate what goes into the box at that location. In this case that there will be three switches located at SSS.

157

| | | |
|---|---|---|
| ◯ LIGHTING OUTLET | | SPECIAL-PURPOSE OUTLET. USE SUB-SCRIPT LETTERS TO INDICATE FUNCTION DW DISHWASHER, DC CLOTHES DRYER, ETC. |
| ⌐◯⌐ CEILING LIGHTING OUTLET FOR RECESSED FIXTURE (OUTLINE SHOWS SHAPE OF FIXTURE) | | |
| CONTINUOUS WIREWAY FOR FLUORESCENT LIGHTING ON CEILING, IN COVES, COR-NICES, ETC. (EXTEND RECTANGLE TO SHOW LENGTH OF INSTALLATION) | S | SINGLE-POLE SWITCH |
| | S₃ | THREE-WAY SWITCH |
| | S₄ | FOUR-WAY SWITCH |
| Ⓛ LIGHTING OUTLET WITH LAMP HOLDER | S_D | AUTOMATIC DOOR SWITCH |
| Ⓛ_PS LIGHTING OUTLET WITH LAMP HOLDER AND PULL SWITCH | S_WP | WEATHERPROOF SWITCH |
| Ⓕ FAN OUTLET | S₂ | DOUBLE-POLE SWITCH |
| Ⓙ JUNCTION BOX | S̲ | SWITCH FOR LOW-VOLTAGE RELAY SYSTEMS |
| Ⓓ DROP-CORD EQUIPPED OUTLET | MS̲ | MASTER SWITCH FOR LOW-VOLTAGE RELAY SYSTEMS |
| —Ⓒ CLOCK OUTLET | | |
| ⊜ DUPLEX OUTLET | [•] | PUSH BUTTON |
| TRIPLEX OUTLET (SUBSTITUTE OTHER NUMBERS FOR OTHER VARIATIONS IN NUMBER OF PLUG POSI-TIONS.) | ╲▯ | BUZZER |
| | ◁▯ | BELL |
| GAS-LINE CONNECTION OUTLET. | ◁▯╱ | COMBINATION BELL-BUZZER |
| | CH | CHIME |
| ⊜_GR DUPLEX OUTLET FOR GROUNDING-TYPE PLUGS | D | ELECTRIC DOOR OPENER |
| ⊜_WP WEATHERPROOF OUTLET | M | MAID'S SIGNAL PLUG |
| MULTI-OUTLET ASSEMBLY (EXTEND AR-ROWS TO LIMITS OF INSTALLATION. USE APPROPRIATE SYMBOL TO INDICATE TYPE OF OUTLET. ALSO INDICATE SPACING OF OUTLETS AS X INCHES.) | ▢ | INTERCONNECTION BOX BELL-RINGING TRANSFORMER |
| | ◤ | OUTSIDE TELEPHONE |
| ⊜_S COMBINATION SWITCH AND OUTLET | ◁ | INTERCONNECTING TELEPHONE |
| ⊙ FLOOR OUTLET | TV | TELEVISION OUTLET |
| —◯—_R RANGE OUTLETS | ▨▨▨▨ | SERVICE PANEL |
| | ▬▬▬▬ | DISTRIBUTION PANEL |
| | – – – – | SWITCH LEG INDICATION. CONNECTS OUT-LETS WITH CONTROL POINTS. |

Fig. 5-6. Electrical symbols.

## RELAY OPERATION

When considering switches and switching, we need to think about the use of a *relay* for this purpose. Relays, which are electrically operated switches, can control the flow of a large amount of power. Just a small amount of power is needed to operate them. They are vital in automatic control systems such as heating and cooling systems. Also, they are used in some other types of remote applications such as an automatic water sprinkler system which comes on at a certain time and goes off at another time. A solenoid type relay may be used to control the water.

A relay basically looks like Fig. 5-8. Figure 5-8A shows, basic relay construction. Figure 5-8B is a single pole, single throw

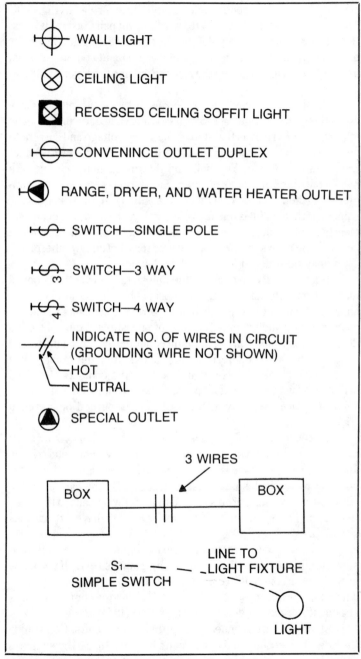

WALL LIGHT

CEILING LIGHT

RECESSED CEILING SOFFIT LIGHT

CONVENINCE OUTLET DUPLEX

RANGE, DRYER, AND WATER HEATER OUTLET

SWITCH—SINGLE POLE

SWITCH—3 WAY

SWITCH—4 WAY

INDICATE NO. OF WIRES IN CIRCUIT
(GROUNDING WIRE NOT SHOWN)
—HOT
—NEUTRAL

SPECIAL OUTLET

3 WIRES

BOX

BOX

LINE TO
LIGHT FIXTURE

S₁
SIMPLE SWITCH

LIGHT

Fig. 5-7. Common symbol designations.

representation. The double pole, double throw, relay in Fig. 5-8C can connect two lines (a and b) to two different pairs of output lines. In Fig. 5-8D the double pole, single throw relay can connect or disconnect both lines from a sheathed cable input completely from a two-wire circuit following the relay. Examine the legend tables of operation.

These relays may operate on as little as 5 volts, from a computer, to the regular 115 volts 60 hertz house current. The importance of a relay is that it may use a tiny voltage and current to operate or control a very large current and voltage to make something run or operate. You will find that most motor circuits are relay-controlled in some manner. Some relays are sealed completely. When you look at them, you simply see a small plastic square with terminals on it. Other relays have the mechanism clearly visible under plastic protective covers. Still others are "open" type relays. Some relays are waterproof and weatherproof. They may be ordered for special applications.

Relays usually give no trouble unless the device they control malfunctions. In that event, there probably will be an excessive amount of current passing through the relay's contacts. This can cause burning and eventually a loss of operation of the controlled device. Sometimes, dirt or moisture can get in the contacts and speed up this kind of deterioration process. It is important, if you do not have a sealed relay, that you keep the contacts clean and bright. Always disconnect all power when working on an installed relay. Replace the relay at the first sign of burned or corroded contacts.

If a relay "sticks," then the device it operates may stay on or may not come on when it should. You can, if you are electrically knowledgeable and take the proper precautions, "jump" the relays contacts to see if this is a cause of trouble. But be careful. If the device is not working properly, and that was what caused the relay to malfunction, sparks may result and you could have a dangerous situation.

Using a single wire current measuring meter, you can measure the current on the lines going away from the relay. If you know how much current that a relay-operated device is supposed to draw when it is working properly, you can tell if the relay or the device is at fault if the device doesn't work as it should.

Electrical wires connected together form circuits. Controlling the flow of electricity in these circuits is accomplished by switches and relays. There are innumerable types of switches and relays.

You can always find one which will fit into any type circuit you can conceive which needs such a unit.

## BRANCH CIRCUITS

The *branch circuit* is the basic circuit for home, apartment and business establishment wiring. A branch circuit is a line which runs from the fuse (or circuit breaker panel) to a single outlet, device, or

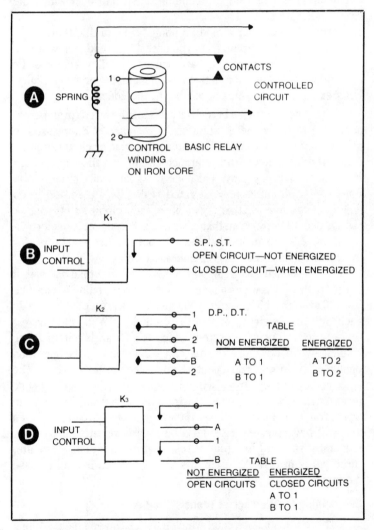

Fig. 5-8. Some types of relays and symbology used. (A) Basic relay construction. (B) Single pole, single throw representation. (C) Double pole, double throw representation. (E) Double pole, single throw designation.

many devices such as receptacles and lights. We have said that a line (sheathed cable, conduit or armored cable containing at least one white wire, and one black wire and a ground wire) which goes from an outlet to the panel box where fuses or circuit breakers are located travels from the device to "home." This line is a "Home run" line. "Home" in this case is said to be the fuse box, switch, or circuit breaker panel on the building or structure in which the circuit is located.

When we consider a modern home, we find that there are a multitude of branch circuits, As you probably already know, these are not put in the home in a haphazard manner. Follow very specific code rules and accepted practices when installing branch circuits. Let's examine some rules according to the code.

The code says that a branch circuit must be rated (in amperes) according to the ampere rating of the *overcurrent device* used to protect that branch circuit. That means the fuse or circuit breaker connected in series with the branch circuit and the supply lines from the electric company. The code says that branch circuits with more than one outlet may be rated at 15, 20, 30 and up to 50 amperes. Of course, there will be branch circuits with higher ratings depending on what that branch circuit connects the electricity to in order to make that device operate. Large motors, for example, require very large amperage ratings.

When you have a circuit which involves more than one pair of wires, such as two wires above ground and one common wire like in a 220 system which is split, you must have a protective device in each "leg" of the circuit except the ground leg. You might have a three-wire or four-wire system which is a "branch" circuit. One example would be where you must have 220 volts for the operation of an electric motor, and a neutral would require three wires. The color coding of these three wires is important. The code says one hot wire must be black, one hot wire must be red and the common or neutral wire must be white. If there is a four-wire circuit, then you will have these wires and a blue wire for the fourth wire. Generally you will find four wires used in three-phase circuits, three wires in two-phase circuits and two wires in single-phase circuits.

## Determining the Number of Branch Circuits

There are some circuits which must be run and have nothing connected to them but the device they operate. Such circuits are for ranges, air conditioners, garbage disposals, dishwashers,

| LOAD | VOLTAGE | CIRCUIT BREAKER RATING | WIRE | | | RECEPTACLE |
| | | | TYPE | SIZE | NO. OF COND'S | AMP |
|---|---|---|---|---|---|---|
| RANGE | 115/230 | 50 | SE | #6 | 3 | 50 MAY DE DIRECT CONNECTION |
| COUNTER TOP COOKING UNIT | 115/230 | 40 | SE | #8 | 3 | DIRECT CONNECTION |
| OVEN UNIT | 115/230 | 40 | SE | #8 | 3 | DIRECT CONNECTION |
| DRYER | 115/230 | 40 | SE | #8 | 3 | 50 |
| WATER HEATER | 115/230 | 40 | NM | #8 | 2 + GND | DIRECT CONNECTION |
| DISH WASHER | 115 | 20 | NM | #12 | 2 + GND | DIRECT CONNECTION |
| DISPOSAL | 115 | 20 | NM | #12 | 2 + GND | DIRECT CONNECTION |
| LAUNDRY (WASHER) | 115 | 20 | NM | #12 | 2+ GND | 20 |

ovens, and most any outlet which will be connected to a device that will draw around 10 amperes (1,000 watts). Freezers and heating circuits are among the types which have a single branch devoted entirely to themselves. Motors which are rated at ½ horsepower or more and blower and oil burner motors need a separate branch circuit. Each of these branch circuits will be separately fused or protected by a circuit breaker, whose rating will be slightly higher than the device which is to be operated. In motor circuits, especially, the fusing must be high enough so the fuse won't blow or the breaker won't trip when the motor starts. Table 5-1 lists some items with the voltages used and the current limitations of the circuit breakers.

Many of these items are directly connected to the branch circuit through a terminal box. A receptacle is not used. Also, in the kitchen area, it is necessary to have at least two 20 ampere outlets for the operation of the various appliances used there. These outlets can be above the counter area, or conveniently located in the breakfast room area if this is adjacent to the kitchen. The distance between such outlets is 12 feet or less, so that a 6 foot cord on a device can easily reach from one to another if the device is centrally located.

There will be a number of single outlet or single device circuits which run from the fuse panel to the kitchen area. If you happen to have a workshop in a basement, attic, garage or small adjacent building, it will have a number of high amperage branch circuits depending, of course, on what you plan to connect to the lines in these areas. For a drill press and lathe, you need lines for each if the motors are reasonably large. If you just have lighting and general convenience outlets, then probably one branch circuit line will be satisfactory providing it has a capacity of 20 to 30 amperes.

Sometimes it will be better to have a separate feed from your house electrical service panel to the shop. There you should have a subpanel with several circuit breakers to fuse and protect the separate branch circuits inside the shop, regardless of where it might be located. The effort in installing this subpanel might be well worth your time and expense.

Now we come to the branch circuits for the rest of the house, condo or apartment. There are several ways to determine how many branch circuits to use. First, multiply the square footage of the home by 3 watts per square foot. This will tell you how many branch circuits to use. As an example, consider a home with 1800 square feet. Making the multiplication we get 5400 watts. Next we divide this by the level of voltage used, 115 volts ac.

$$\frac{5400 \text{ watts}}{115 \text{ volts}} = 46.9565 \text{ amperes}$$

Dividing the amperes by 15 amperes per circuit, we get three branch circuits (practically), although we could use four branch circuits if we aren't interested in cost. An electrical installation company would probably use three branch circuits in this case.

The method determines how many branch circuits to use for the rest of the house after you have considered the number of branch circuits for the kitchen and other special demand devices to be used in the home. When you consider that the general accepted practice in home electrical design is to have 1.5 amperes per outlet, that means that you can use up to eight outlets on a branch. As one professional engineer told us, you can use up to eight devices per branch, including lights and receptacles in home, condo or apartment wiring.

You must be aware that the branch circuits devices (outlets or receptacles) must be of such a rating that they can handle, individually, the total current for which the branch is designed. If the branch is rated at 15 amperes and is protected by a circuit breaker of that rating, then *the wiring and the receptacle* (outlet) must be rated at least 15 amperes. Of course, it is planned that no branch will have over 80 percent of its capability placed on it at any one time. A 15 ampere branch circuit will never have to supply more than 12 amperes when fully loaded with a family's demand.

### Tips on Branch Circuit Wiring

We considered the amount and type of electrical equipment which might be found in a home and questioned many knowledgeable persons about the situation where much of this equipment might be placed on a single branch. The experts say never wire one room with a single branch. The common practice is to wire a single room with anywhere from two to three branches. The reason for this is to prevent overloading of a single branch in a single room and provide for lighting if one branch blows out a fuse or trips a circuit breaker. Plan your wiring so that you do not have all outlets and lights in a room on a single branch wiring circuit. Use several branches if possible and, as a minimum, have at least two branches.

It is possible for a branch, even if it is split and supplies two or more rooms, to become overloaded. The breaker will normally trip when the circuit is overloaded. You really have to worry if the protection device, be it fuse, circuit breaker or whatever, does not operate. Then you have a problem. Do not economize or get cheap protection equipment. Get the best. Don't wire around fuses or breakers. Be sure that your circuit breakers are the right size, wired in properly, and are electrically perfect.

### TWO-ROOM WIRING DIAGRAM

Figure 5-9 is a simple sketch showing how some branch circuits are used to wire lights and receptacles. We have included the dotted lines to show the connecting cables which have two wires and ground (14/2). You can see in this simple example how the "takeoff" for the "homerun" is indicated. The "home" is the fuse or circuit-breaker panel.

Four basic circuits are shown. In the large room at the right of Fig. 5-9 we have only four receptacles on each circuit branch. An electrical contractor might put all eight of them on one branch circuit. We show four because, if this is a family room, you do not want all circuits on one branch. You might want a heavier loading on the circuits for the devices in that room. Lights are always on their own circuit. Additional lights in other rooms might be connected to this branch circuit.

Black and white connecting wires are used in the cables, and the white wire is connected to all other white wires. The bare or green grounding wires are all connected together. Lines from switches are black and blue or red. Look again at the symbols in Figs. 5-1 and 5-2 and the specifications in Table 5-1. You will use this data when you do your practice plan later on.

It seems pretty clear that, aside from knowing the number of wires in a cable for a given branch, you could layout your home's electrical systems, specify where you want outlets and lights, and specify the kind of lighting (fluorescent or incandescent) to be placed in each location. The only thing you have to remember is that the code says you cannot connect the outlets in the kitchen to any other outlets or lights in other rooms, except perhaps the laundry room. The code also tells you that the outlets must be so spaced that a maximum of 12 feet linear distance between them is met. They cannot be further apart than that.

When we examine many of the modern homes and plans for them, we find that separate lighting circuit branches are used that are not connected to receptacle circuits. Separate circuits are used for fixed appliances such as stoves, ranges and ovens. There should be only eight devices on a branch according to accepted practice. There probably should be at least one spare electrical circuit installed for every five circuits used, just in case.

City ordinances may dictate the type of conduit used. You should check out the ordinances for your area to find out what restrictions, if any, are placed on electrical wiring by them.

## BRANCH OVERCURRENT PROTECTION

There is a requirement that the fuse or circuit breaker disconnect panel, whatever type it may be, must be located near the point where the wiring enters the building. Since this location is the "home" of the "home runs" of the branch circuits, it probably will be designated with the assistance of the electric company. The electric company will tell you where they will locate their transformer or line feed service pole.

Every branch and all circuits within a home, business or shop must have overcurrent protection. Branch circuits to receptacles are rated at 15 or 20 amperes. The rating to fixed appliances and devices may be higher, such as for the air conditioner. Lights are normally on 15 ampere branch circuits. Any outside branch circuits may have to have special overcurrent and ground fault indicator protection, as might any circuits in shops on the home premises. Since we have mentioned ground fault indicators earlier, let us examine them and their installation in more detail.

First we will examine in Fig. 5-10 the exit of "home run" wires in a home under construction. The circuit breaker panel and electric meter must be located close to this point. Each pair of these wires will connect into a circuit breaker panel.

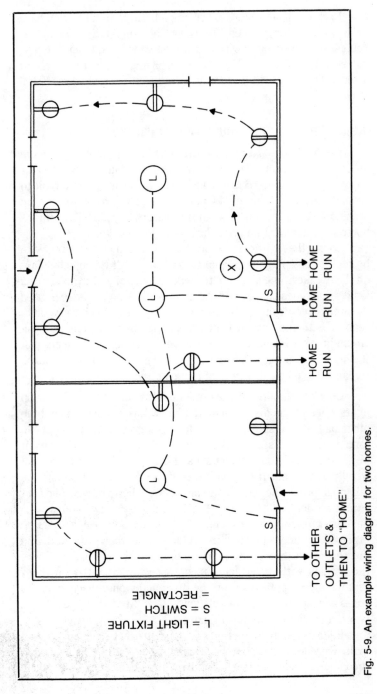

L = LIGHT FIXTURE
S = SWITCH
= RECTANGLE

TO OTHER
OUTLETS &
THEN TO "HOME"

HOME
RUN

HOME
RUN

HOME
RUN

HOME
RUN

Fig. 5-9. An example wiring diagram for two homes.

Each of these cables represents a branch circuit as you have probably already guessed. There is one single wire among the cables. It is the ground wire and will be connected to an iron pipe driven at least 6 feet into the earth and having a diameter of at least 1 inch. The wire will be tightly bolted to it for a good electrical connection.

## INSTALLATION OF THE GROUND FAULT INDICATOR

You have to isolate the circuit in which you want to place the ground fault indicator. Remember that the ground fault indicator trips a breaker if there is any leakage current from the circuit to ground. Normally this might not be enough current to trip the breaker directly, but may constitute a hazard to people living in the home. The installation procedure is as follows.

Locate the fuse or circuit breaker controlling the receptacle to be replaced by plugging a lamp or radio into that receptacle. Turn off the breakers one at a time to remove fuses one at a time till the lamp's light goes out or the radio goes dead. Now you have located the breaker controlling that particular receptacle which you want to replace with the ground fault receptacle. If it happens that the circuit breaker is a double pole type on that circuit, this means that the branch circuit is a multi-wire type. The ground fault receptacle cannot be used on this type of branch circuit.

Check both outlet halves of all duplex receptacles. If a lamp lights when plugged into one outlet half of a receptacle and not the other half, it is a split wired circuit. The ground fault receptacle cannot be used as a replacement for this type of arrangement because the individual receptacles of the split wired unit are electrically separated. If one half is controlled by a switch, and the other half is not, the receptacle cannot be replaced by a ground fault receptacle unit.

After finding those receptacles on a circuit which could be replaced by the ground fault receptacles, and thus protected, find the receptacle nearest the circuit breaker or fuse panel. Check receptacles with a test lamp to be sure the circuit is off. Now remove the receptacles from the box one by one, disconnect the wires and tape the ends with electrical tape. Mark the wires so that you can replace them exactly later.

Feed through protection of "downstream" receptacles on a multi-wire circuit cannot be accomplished with a ground fault receptacle. If there are no multi-wires, locate the receptacle

Fig. 5-10. "Home run" wires exiting a house.

nearest to the fuse or switch box. When you find the nearest one to the fuse or circuit breaker box and disconnect its wires, then all other receptacles "downstream" from this one, and connected to it, will not light a test lamp when power is turned on. Examine Fig. 5-9 and notice the (X), which is the receptacle nearest the circuit breaker panel on that particular branch circuit. The arrows in the dotted lines indicate that all lines are in parallel with the (X)

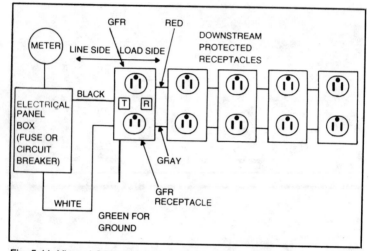

Fig. 5-11. View of GFR receptacle and wiring connections for a feed-through installation.

receptacle. When the lines are opened at (X), the other three receptacles go "dead."

If you disconnect a receptacle and the test lamp, when plugged into other receptacles, lights, then you may have disconnected a "downstream" receptacle. You then have to put this back together and try another until you find the one nearest the breaker panel.

When you find the receptacle nearest the fuse or breaker panel, replace it with the ground fault receptacle, following the wiring instructions with it. You will gain ground fault protection over all receptacles on that branch. It is wise to label this receptacle so that it can be identified easily.

## WIRING INSTRUCTIONS AND RATINGS
## FOR THE BRYANT GFR PROTECTION UNIT

It is always helpful to examine some electrical specifications and look at some wiring diagrams of a unit like the Bryant GFR receptacle. Do this to understand what the unit is, how it works, how to connect it into a branch circuit and how to test it.

The Bryant GFR receptacle is designed to provide reliable protection against shock hazards of line-to-ground faults. When it is properly installed, the GFR device automatically interrupts the circuit when it detects low level current leakage in excess of 6 milliamperes. Thus it protects persons against injury or electrocution. It is a UL Group I device that conforms to UL Standard 943.

170

The receptacle itself conforms to the UL Standard 498. There are two types of these units. The GFR 52 is ground fault rated for 15 amperes/120 volts 60 hertz ac, and the GFR 53 is ground fault rated at 20 amperes 120 volts, 60 hertz ac. Remember that the devices will trip the breaker on any current in excess of 6 milliamperes which is due to a ground fault leakage.

The installation diagram for this kind of protection unit is shown in Fig. 5-11. In Fig. 5-12 we see the back of the unit and the

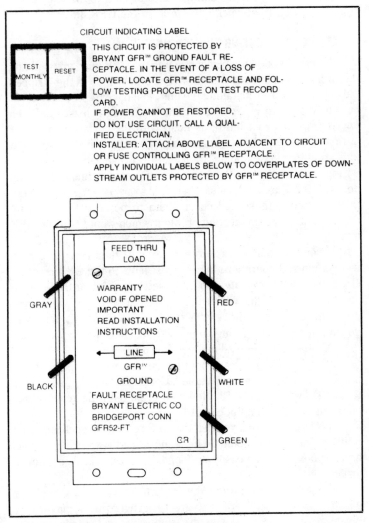

Fig. 5-12. Wire designation to a GFR receptacle.

wiring connections to it. Notice particularly the colors of the wires used in the diagrams of Figs. 5-11 and 5-12. We find red and gray wires needed in addition to the normal white, black and green (or bare) wires coming from the circuit breaker panel. The red and gray wires connect the downstream receptacles in a parallel arrangement and are protected because they are fed through the GFR unit. If you use the GFR as the last receptacle on a branch, say as a bathroom outlet unit, then you would not use the red and gray wires. You simply use insulating "nuts" or tape up the ends of the red and black wires which emerge from the GFR unit.

## RESETTING AND TESTING THE GFR UNIT

When you open the box containing the GFR unit, you will find that it has a trip and reset button. The unit will be in the "tripped out" position. After installing the unit, you have to press the reset button till you hear a definite click when you release it. When energized, the reset button is flush with the surface of the test button. If the reset button will not reset, or it continues to pop out indicating that it is tripping, then check downstream for improper wiring, cuts or nicks in the insulation of wires, or moisture in the boxes. If there are appliances connected to any downstream receptacles, remove these and check them to be sure they are not causing a leakage or shorting problem which would trip the GFR unit.

If you find that you can reset the unit, then test for proper installation and operation. Insert a test lamp in the GFR receptacle. Press the *"test monthly"* button (see Fig. 5-12 for labels to be applied to GFR). This will simulate a fault in the circuit. When you press the test button, the reset button will pop out and the lamp will go out if the unit is working properly. If this does not happen, then the unit itself may be bad and need to be replaced.

If the test lamp stays on but the reset button does pop out, the line and the load wires are reversed. Correct the wiring and be sure it is as shown in Fig. 5-12. Now test again until the reset button pops out and the lamp goes out.

It is important that you test the GFR receptacle each month. A written test record should be kept and placed in the circuit panel breaker box so it can easily be found. This will insure maximum protection.

We wish to emphasize the importance of testing a circuit which is protected. There is no greater danger than to have some electrical system fail because the protective device fails. Always test the protection devices, be they smoke detectors, GFR units or whatever, to be *sure* they are working properly.

## TESTING A SMOKE DETECTOR

We have just described a specific testing procedure for the GFR units. *Testing a smoke detector* requires smoke. Do not be misled by advertising which says that the smoke detector has a test button. When you depress the button, you may be testing nothing but the horn alarm circuit. You may not be testing the smoke detection unit. Your life depends on that unit performing properly.

Use a can with holes around the base and in the lid. Fill this with burning pieces of paper which have been dampened slightly. When you have smoke coming from the holes on top, pass it under the smoke detector and let the smoke hit the unit. If the alarm goes off, it is working properly. If the alarm doesn't go off, the smoke may be too thin. Try a little more smoke. Then if it doesn't work, replace the battery, check the electrical connections if it is connected to the house wiring, or buy a new unit. Never rely on just one smoke detector. Get one for each hall, one for each bedroom and one for the kitchen. It isn't that expensive when you compare the price of detectors to the cost of a home.

One other feature you might look for in a smoke detector is a low-battery indicator. Some units sound an alarm when the battery gets low; others may light a light.

## ELECTRICAL WIRING DIAGRAM OF A MODERN HOME

Now that we have the proper background, we will examine a blueprint diagram of the wiring (actually the devices to be installed in a modern home. We show this through the courtesy of Perry Homes of Houston. It represents one of their average size homes. They build both large and small homes along with condominiums and apartments. Look at Fig. 5-13.

In order to be able to lay out the electrical requirements, we need to look this plan in Fig. 5-13 over very carefully. Notice, first, that the actual "runs" of the wiring itself are not specified. We leave this up to the wiring installer. In the next chapter we will examine some illustrations of actual wiring being performed and the "roughing in" processes preceding such wiring. Here we will be examining the way the architect lays out the electrical plan.

Let us examine the house to be wired as shown in Fig. 5-13. It is a four-bedroom home with two baths. It has a family room, a living room, dining room, kitchen, breakfast room and a utility

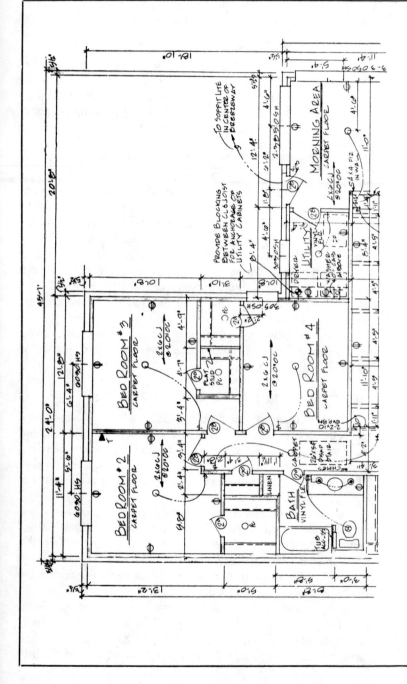

174

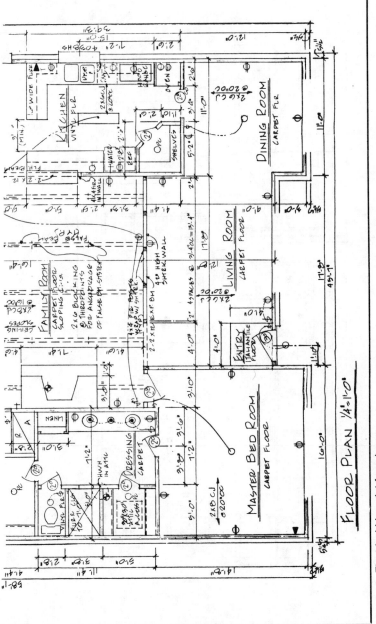

Fig. 5-13. Electrical blueprint for a home.

175

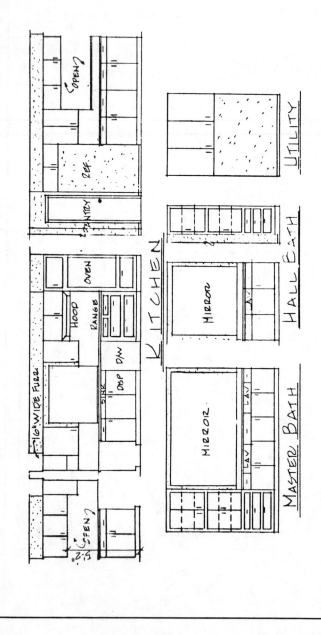

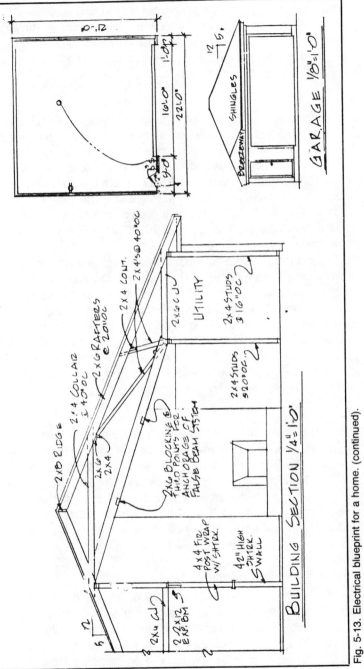

10'-12

1'-0"

10'-0"

22'-0"

4'-0"

12
5

SHINGLES

BREEZEWAY

GARAGE 1/8"=1'-0"

2×8 RIDGE

2×4 COLLAR
@ 4'0OC

2×6 RAFTERS
@ 20"OC

2×6
2×4

2×4 CONT.

2×4's @ 40"OC

2×4 CONT.

2×12 J

UTILITY

2×4 STUDS
@ 16"OC

2×6 BLOCKING @
THILD POINTS FOR
ANCHORAGE OF
FALSE BEAM SYSTEM

2×4 STUDS
@ 20"OC

4×4 FIR
POST W/SHT.RK.

42" HIGH
SHT.RK. WALL

2×4 CJ

2-2×12
EXP. BM.

12
5

BUILDING SECTION 1/4"=1'-0"

Fig. 5-13. Electrical blueprint for a home. (continued).

177

room. The garage is a detached unit, but it must have a light and a switch for the light. It is usual with a detached garage to run an underground line, using proper moisture-resistant cables which are placed inside moistureproof PVC conduit. The conduit has good tight fittings at each end.

## Bedroom # 2

You will next probably want to go back to the symbols in Figs. 5-6 and 5-7. Identify the various electrical outlets, lights and devices connections by means of these symbols. Let us start examining the blueprint at bedroom #2. Look around the walls; you will find an outlet in each wall. The switch for the overhead light and the run from the switch to the light are in the center of the diagram. The switch is near the right-hand door. Notice the filled-in triangle on the right wall marked "T." This is a telephone connection.

As you can see from the outside dimentions of this room, the size is about 13 feet by 11 feet. The placement of a wall receptacle in the center of each wall meets the code requirement that receptacles not be over 12 feet apart, linear distance, or that a lamp with a 6 foot cord can reach an outlet. You may want to refer back to Chapter 2 to refresh your memory. Notice how the light switch is carefully placed on the open side of the door, even though it is just around the wall. You would be able to reach it easily by opening the door just a bit to pass your hand through.

## Bedroom # 3

Now examine the next bedroom to the right—bedroom # 3. Again, judging from the room dimensions, we find outlets, one on each wall. The light switch is near the open side of the door. There is a dotted line showing which light this switch will operate. Next to the closet in this bedroom, near the outer wall, is a small storage room. In its center you see a small circle with the designation $P_c$. This designates a light with a pull chain or pull cord. Actually bedroom # 2 also has such a closet.

## Hallway, Bathroom and Lights

Now move out into the hallway below and near the left-hand corner of bedroom # 4. You find a three-way switch arrangement for the hall light which, incidentally, is shown drawn over the word "carpet." You trace the curved line from the light both ways and find it terminates in a switch marked $S_3$, meaning three-way switch. As you already know, this means the hall light is to be operated from each end of the hall regardless of how the switch at the opposite end is positioned. Using three-way switches and connecting them properly will accomplish this objective.

The hall bathroom, just below bedroom # 2, has an overhead light. There are two switches in the same box near the doorway. The second switch turns lights on over the vanity wash basin. Notice that there is an outlet at the end of the counter for shaving, lights and hair dryers. This particular outlet is often a GFR unit since it is located in the bathroom and would be on its own branch circuit. Below the bathroom and to the left is another closet with a pull chain light.

Now we examine the bathroom area of the master bedroom, which is down from the closet. Notice that there is an overhead light over the commode with a switch near the doors. There are three recessed lights (Fig. 5-14). The switch to control these lights, which are wired in parallel, is near the doorway to the

Fig. 5-14. A recessed light fixture for a bathroom in an unfinished ceiling. A glass cover will be placed over this unit.

bathroom. These lights are over the vanity wash basin counter. At each end of the counter are 110 volt wall receptacles.

Across from the wash basin is another closet with access to the attic. There is a pull chain light in the attic at this point. In Fig. 5-15 we see an example of this kind of fixture with a bell transformer connected to it. The bell transformer's wires on the low voltage side will run to the door bell and the chimes. They are connected in series with this low voltage winding. The pull chain fixture attaches directly to the box of metal or plastic which is placed in the ceiling.

## Air Conditioning and Heat Exchanger Units

The air conditioning unit is located in the attic in most cases. Of course, the heat transfer part of the unit is located outside. The blower and cooling coils and the heating element (if it is a heating-cooling unit) are in the attic. Figure 5-16 shows the attic unit.

The heat exchanger unit, which we often see at the side of a home, is the unit which causes the hot gas to be converted back to a liquid through its compressor. This unit has a large electric fan to draw air through the cooling coils and expel the heat from inside the home. Also, it has a large motor on its compressor. It is normal for a 220 volt connection to be made here. You can see the end of the large cable with its black, white and bare wires not yet connected to this new unit. Look at Fig. 5-17.

While we are on the subject of air conditioning and heating, we will examine in Fig. 5-18 a small unit which is located inside a two-bedroom apartment. The apartment had a living room, bath and small kitchen. There was a small closet area in which this heating-cooling unit was located. It is in an unfinished wall. If the wall is finished and painted, you would see only a grill air intake and output.

The heat exchanger should still be located outside the building. Since this apartment was in a building several stories high, there was no attic for locating the blower and heating unit. So it was located right in the apartment's special "closet."

## Overhead Lights In Room Centers

Return to the blueprint in Fig. 5-13. Look at the master bedroom and you find, again, the receptacles in the walls and the telephone outlet location. This type of telephone symbol is for an outside line connection as differentiated from the internal tele-the kitchen proper. That light is controlled by one of three

Fig. 5-15. A pull chain fixture with a bell transformer connected to it. Notice the black input wires for 110 volt line and the smaller wires for the 6-8 volt output lines from the transformer.

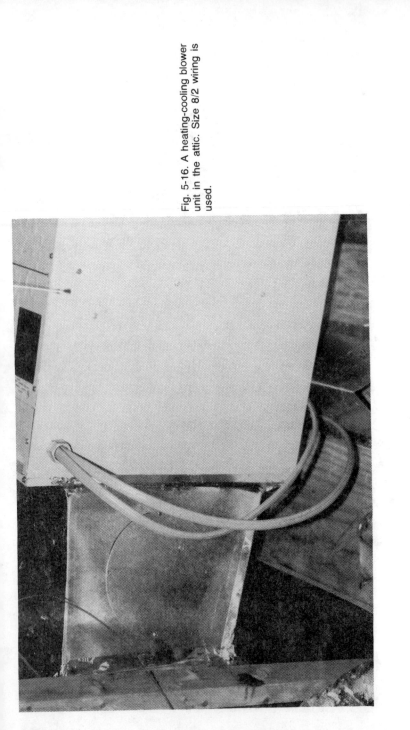

Fig. 5-16. A heating-cooling blower unit in the attic. Size 8/2 wiring is used.

Fig. 5-17. The heat exchanger unit of the air conditioner system is located outside the house.

183

phone system which some larger homes may have. Notice the "run" from the light switch at the door to the over head light in the center of the room.

Some homes which were built many years ago do not have these overhead lights in the centers of rooms or anywhere else in the room for that matter. In modernizing some homes, it is necessary to run some cable to install such lighting. If you have attic access, the overhead part is easy. You may tap into some attic box for a source of voltage. Running the cable down the wall near a doorway may be difficult. You may have to use surface wiring to do a good job. If you plan a new home, it is always better have this overhead lighting. If you don't use it, that is fine. You can actually take off the fixture and seal over the electrical box so it won't show. But it is there if you ever need it, and the installed switch is not a bit conspicuous. It's better to be over-wired than not to have enough branches, outlets and fixtures in the home.

## Utility Room

A quick glance at bedroom # 4 shows us that it is wired like the others. We move to the right to the utility room. Here we find the symbol for a gas outlet for the dryer, and electrical outlets for the washer and freezer (DF). Each of these electrical "runs" will be on a separate cable and have its own breaker switch in the fuse breaker panel. Sometimes we find the electrical service meter connecting point just outside this utility room, but that will depend on the location of the electric company's utility pole or underground connection location. Short runs from high amperage outlet locations to the circuit breaker panel are desirable, if possible.

## Three Important Switches

Moving further to the right we examine the morning area and high amperage receptacles must be located there, too. The receptacles should have 20 amperes each and be connected on their own cables to the breaker panel. This is done so you can operate various appliances in the area without blowing a fuse or tripping a breaker. Notice that here is a case where a switch will operate an outside light in the breezeway to the garage. That run must be made in moistureproof or moisture-resistant conduit or cable. It will be found that the breezeway will have a roof, and the wiring will be protected under it also. Near the kitchen is another light, which is actually over a counter between the morning or breakfast room and

Fig. 5-18. A small internal air conditioning unit.

switches, all in the same box, near the doorway to the family room. When the wiring is installed in the building stage, we find that this particular switch looks like that shown in Fig. 5-19.

These three switches will control lights in the morning area, the kitchen and in the family room through a wall receptacle. Notice also that the wall receptacle is on a three-way switch to the doorway at the bottom left corner of the family room (Fig. 5-13). Thus, a lamp connected into this receptacle can be turned off or on from either the kitchen doorway or the master bedroom doorway. The wiring of these switches, lights and receptacle will require some thought on the part of the electrician who does the job.

## Fireplace Area

Notice also that there is a 110 volt outlet near the fireplace area. This is a convenience feature for electrical lighting or whatever. There is a gas outlet in the fireplace for connection to a gas burner in order to light a wood fire easily. Above the fireplace is the hall opening which has no doorway. Below the fireplace and to the right of the master bedroom doorway is an opening to the family room from the living room. It does not have a doorway either. This opening is in the corner (upper left) of the living room. It is not too clearly shown.

## Living Room

The living room does not have an overhead light, but it does have a receptacle connected to a switch near the entrance doorway. There is also a switch for the outside light in the entrance. Since the entry goes into the space for the living room, we find no outlets on the left-hand wall. Because the living room is constructed the way it is, we find few outlets around its perimeter. There is an open wall to the dining room to the right. There we find an overhead light with a switch near the kitchen doorway. There are outlets on most walls of this area.

## Kitchen

We have left the kitchen until last. There we will find all those outlets which the code required. Many of the appliances will be directly wired into boxes. Counters will have 20 ampere receptacles separately fused as the code requires.

## Electrical Contractor's Duties

This blueprint in Fig. 5-13 is obtained by the electrical contractor from the builder, who gets it from the architect who designs the home. It is up to the electrical contractor to see that all outlets, lights and electrical connections are installed in the best and most professional manner. He has to decide how and where to make the cable runs and he determines the size cables and wiring to be used to meet the National Electrical Code requirements. There may also be city ordinances which govern the electrical wiring.

Fig. 5-19. Three switches will occupy this box. Notice the three Ss stating this fact. White wires are connected together, as are all bare wires.

You have probably already noted that the blueprint does not show the electrical installation for the air conditioner and heating system. Nor does it specify the wiring for doorbells or any security wiring. These, of course, need to be included.

## DESIGNING YOUR HOME'S ELECTRICAL INSTALLATION

Since it is always nice to do-it-yourself, we now include what some might call practical exercises in the design and layout of electrical installations in homes. In these days of economy, it is often possible to obtain some land and have a contractor put up a shell or framework for your home. Then you finish the job. You save literally thousands of dollars, and all it takes is a little time, patience and ingenuity. The whole family can get involved in the finishing of a home.

### Using Plastic To Make a Working Cover

With the background you now have from the previous chapters in this book and the detailed examination of the blueprint of a home, you should now be able to take the home layouts as shown in Figs. 5-20, 5-21 and 5-22, and draw in the electrical installation you want. Make sure it meets the code requirements. We suggest that you get a small piece of clear plastic and lay it over each illustra-

Fig. 5-20. A layout for a three bedroom home with a breakfast room.

tion. Then draw on the plastic with a crayon or grease pencil for the first time or two. This way makes it easy to change things. These homes in Figs. 5-20 through 5-22 are standard types. If you are very progressive, you might take these drawings and expand or contract them to meet your needs before you do the electrical installation design. Look now at Figs. 5-20 through 5-22.

When you use plastic to make a "working cover" for the drawings, you can also indicate to yourself how you might plan the "runs" of electrical cables to a "home" or fuse panel circuit breaker location. You will remember that in the best electrical design you will not have over eight devices on a single branch. You will recall that many of the kitchen outlets, devices and appliances will require their own circuits. And you can use three-way switches to make the light control throughout the home easier.

## Planning a Workbench Area and Outside Lighting

In cases where the garage is attached, you might want to plan a workbench area. Be sure to have enough electrical amperage capability and outlets to operate all motor-operated devices such as saws and drills which you will use there. Then, too, you will surely want adequate lighting. You might specify fluorescent lights over the work bench.

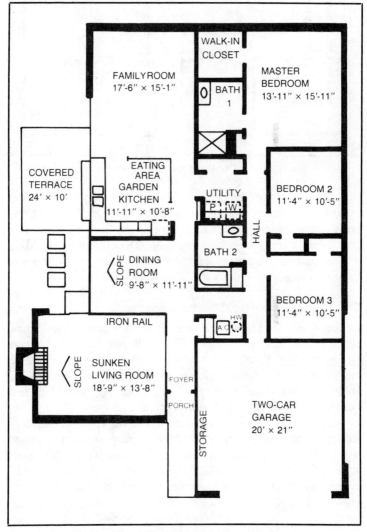

Fig. 5-21. Layout for a three bedroom home with a sunken living room.

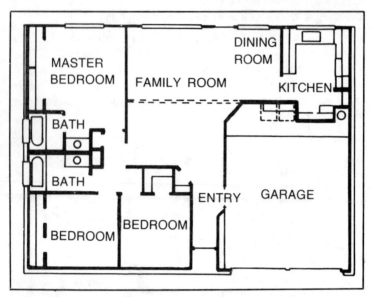

Fig. 5-22. Layout for a three bedroom home with a minimum of "extras."

If you want the outside lighted at night, you might plan for the underground runs of waterproof type cables, conduit or combinations to the various points where lighting might be installed. You may want a central switch area to control this illumination. Also, you may desire weatherproof outlets on the outside of the house at various points to operate such devices as grass trimmers.

### Determining Service Input Requirements

You will need to know how to determine your service input requirements from the electric company. We suggest that after you have completed your plans, stop by the electric company's office and talk to their consultants. They may point out something you've forgotten or overlooked, and they surely can suggest the type and size of input service you would need.

You may not want to try to do the wiring yourself. You will have saved money already by doing the planning, so you might want to contact a master electrician who can do the job for you. If you are an electronics or electricity buff who has some experience, you will have some "feel" for the work and probably can do it properly. However, if you have any doubts about your ability, get a master electrician to do the job for you. If you furnish all material and he only supplies the labor, the cost will not be as great.

# How To Wire A Home  6

According to a study by the government, there are two separate phases of installation which are performed by master electricians on new residential construction. These phases are the *rough-in work* and the *finish work*. The master or journeyman electrician does the work, assisted by helpers who often are apprentices learning the trade. Apprentices have a requisite understanding of electrical wiring and need the experience of "on the job" operations for four years to become journeymen electricians. They start when the interior and exterior walls have been framed in, even as the slabs are poured, in some cases, where conduit must pass in or through the cement. The job is finished when they make the final installation of switches, lights and connections to ovens, ranges, etc., after the walls have been finished.

In Fig. 6-1 we see the outside wall of an apartment being finished. The interior is still in the framing stage and looks like Fig. 6-2. Homes still will also look like this and, except for size and number of units, will be handled the same way. Condominums are just apartment type units which are sold and not rented.

## ROUGHING-IN PHASE

The master electrician starts his work with the arrangement shown in Fig. 6-2, the rough-in. Now he will use the blueprints. The supervisory electrician will locate the spot where the utility pole conductors are tied to the building or where the underground

Fig. 6-1. The outside wall of an apartment is being finished.

system transformer will be located. Refer again to Fig. 3-1. From this location he will determine where the "home run" wiring will exit from the building wall to be later connected into the fuse or circuit breaker panel and electric metering system. It is the electrician's responsibility to assemble and connect all conduit, weatherhead and meter loop and all interior circuit breaker panels. The panel(s) will be the heart of all the interior and exterior wiring of the building. It is the "home" of all "home run" wiring.

Next the rough-in phase continues with the installation of the receptacle and light and switch boxes. These may be plastic or metal. If the wiring is performed in Romex, which is the technical name for the nonmetallic sheathed plastic cable we have referred to in previous chapters (a two wire plus ground cable), the plastic boxes might be used. If metal conduit is used, metal boxes probably will be used as the fittings for cable ends and box connection are designed for this relationship. Plastic Romex cable may also have fittings if this cable goes into a metal box. When the cable goes into a plastic box, such fittings are usually not used.

Plastic light fixture boxes can be simply fastened to a roof or ceiling support member as shown in Fig. 6-3 by hammering down the two nails on each side. The metal water pipe behind the box is far enough from it so the two do not touch. By the way, the location of the boxes which we are examining is determined from the

Fig. 6-2. Inside of a home under construction. The walls have been framed on a cement slab and the roof is in place.

Fig. 6-3. Installation of a light fixture box during rough-in. This plastic box has two nails which are driven into the support structure for the mounting.

blueprint. You now see the importance of specifying, definitely, all outlets, switch and light fixture locations on that drawing.

When you can nail to a beam or 2×4 as shown in Fig. 6-3, you are in good shape. Often the blueprint will call for a light box, receptacle or a switch to be located between joists, studs or the wooden structure of the framing setup. In this case you use hangers, some of which are shown attached to light boxes in Fig. 6-4. These hangers will be nailed to the beams or studs and the box moved to its desired position. Then it is tightened in place by tightening the bolt or plate on its back. The two types of mountings for light fixtures of the common or regular type and the boxes are shown in Fig. 6-5. The near box uses the nails on the side; the upper one uses a linear hanger.

Now let us examine some installations of boxes with these hangers. In Fig. 6-6 we see how the location of a box between two plastic pipes is accomplished. Notice that the ends of the hanger have been bent so that the hanger itself is recessed and flush with the line of the studs. Another view of how an overhead light fixture is centered between two beams is shown in Fig. 6-7.

### Recessed Lighting

In some areas like the bathroom and the kitchen, you will probably find recessed lighting which uses a higher intensity light.

Fig. 6-4. These hangers help to place boxes in precise locations.

Fig. 6-5. Two mounting arrangements are shown. The boxes are deep and shallow types. A hanger is used for one box and nail mounting for the other.

This requires a protected fixture such as shown in Fig. 6-8. Notice how two hangers are used to position this kind of fixture exactly where the blueprint calls for it to be.

These kinds of fixtures will use a flat translucent glass plate over them to diffuse the light. They will also use a higher wattage bulb than most light fixtures, say from 150 to 250 watts or even higher. Another view of this type of fixture is shown in Fig. 6-9. The electrical supervisor goes around marking on studs or beams where and what type of unit will be located. A box for switches is shown in Fig. 6-10.

### Switch Box and Receptacles

Notice that the supervisor has designated the bottom level of the box by the two connecting lines. He has indicated how many switches are to be located at this point. When the helper or apprentice (who usually actually fastens the boxes in place) looks at

Fig. 6-6. A hanger attached to studs places the box between plastic pipes.

the code information on the stud, he selects the right box size for two switches, not for three or one. He also makes sure the box faces the right direction, toward the letters, and at the right height.

Sometimes it isn't easy to place a box in a given location. In Fig. 6-11 the box had to be located near the doorway studs. It was

Fig. 6-7. A hanger is used on ceiling joists to position for a light fixture.

Fig. 6-8. Two hangers are used to position a recessed light fixture.

necessary to cut out a piece of one stud to fit the box in place. Fortunately, there is room between the two pair of studs for the wiring when it is later installed. Ask yourself how you would hammer those nails. You can't. In this case we were told that the electrician also had to take down the right-hand studs to install the box and then put them back in place. If you want to become an electrician, be aware you might also have to be somewhat of a carpenter and a plumber for rigid conduit.

As we look around at other receptacles placed during this roughing-in phase, we see some placed back to back as in Figs. 6-12 and 6-13. Be aware, however, that this does not mean they will be wired in parallel on the same circuit. Since some boxes may have receptacles which are operated from a switch, they may be on two entirely different circuits although they are back-to-back. The electrical supervisor has to determine if this is the case. He also specifies how the lines will be routed and just how everything will be connected. Examine Figs. 6-12 and 6-13 showing the boxes about 18 inches above the floor level.

Finally we show a switch box and a receptacle as they have been roughed-in on adjacent studs (Fig. 6-14). Imagine yourself walking through the open studs of your own home locating the positions of the various boxes according to your own blueprint and fastening them in place. You will, at the same time, be considering how many devices will be on a circuit, and how the Romex will be

Fig. 6-9. A side view of the recessed light fixture.

199

Fig. 6-10. Shown is a box for two switches.

Fig. 6-11. Installation of a box is not always easy.

Fig. 6-12. Two back-to-back boxes on adjacent studs. These may not be wired on the same branch circuit.

Fig. 6-13. Two boxes are located back-to-back on back-to-back studs. These also may not have the same branch circuit feeding them.

routed to connect to all of them and then go directly to the circuit breaker panel location. We now explore the wiring in great detail.

## ROMEX CABLE

Romex or non-metallic sheathed cable is normally referred to by the size of the wires and the number of wires that are wrapped together in a plastic cover. A 12-2 with ground means that there are two insulated wires or conductors (white and black insulation), each with a size 12 wire inside the insulation. There is also the bare or green ground wire in the cable.

## DETERMINING WIRE RUNS

All circuits originate at the circuit breaker panel and spread from there throughout the house. After the rough-in phase, all wiring must be completed before the walls and ceilings are insulated and covered with gypsum wallboard. Therefore the master electrician and his helpers must work rapidly, but exactly, to complete the wiring. This does not include the installation of switches, lights and receptacles. It does mean the wiring must go from the boxes to the circuit breaker panel and be ready for the installation of these devices after the wallboard has been put in place.

The electrical wiring normally is subject to an inspection by the city inspector after it has been completed. It is very important that the wiring be done correctly. If mistakes are found after the wallboard is in place, then it can be very costly to tear down the wall, fix the mistake, and replace the wallboard. Errors must be corrected to the inspector's satisfaction.

Now that we have the boxes in place and a circuit breaker panel location, we are ready to put in the wiring itself. The electrical contractor should determine the best and shortest routes for each branch circuit. This saves money in material costs. It is important that the proper "runs" of wire be carefully considered, not just for the monetary savings but also for the more efficient use of electric current. Even though copper wire is a good electrical conductor, it does offer some resistance to the flow of electricity. The longer the wire, the more resistance will occur and the less voltage will appear at the far end of the "run." Short runs are vital when you plan large current drain branches.

## DRILLING HOLES IN BEAMS

As we see in Fig. 6-15, the next step after deciding where the wires will run will be to drill holes in the studs and beams to make it possible for the wires to take the routes chosen. The wire will

Fig. 6-14. Rough-in of a switch box and receptacle box on adjacent studs. The horizontal "V" shows the height at which the box is to be set.

Fig. 6-15. When cable routing has been determined, then it is time to drill holes for the "runs."

come from overhead, down the 2×4 which is being drilled and into the outlet box below. The holes must be in the center of the 2×4 studs so there is at least a 1½ inch depth before a nail could reach the wire. In cases where it might be necessary to notch a stud or beam to run wires, that notch must be covered by a steel plate so a nail cannot penetrate. The hole is large, usually about ¾ to an inch in diameter, so the wires run and pull easily through them even if the run is quite long.

As shown in Fig. 6-15, there is a simple straight run. You may be wondering how to get around corners. Figure 6-16 will show you the answer. The run here comes from a right angle to the line of the drill bit. The bit pierces both the stud facing and the one sideways to the run so the wire can pass around the corner when the hole is complete. Sometimes it is necessary to drill from two or even three directions to get a route through the studs (or beams) to get the runs going where they should. A careful study of Fig. 6-16 shows that the entrance of the drill was at an angle to the plane of the facing stud. The long drill is necessary to reach "out of the way" places.

## CONDUIT

If electrical conduit is used, it is much more difficult to run since rigid pipe, be it plastic or metal, just doesn't have the flexibility of the Romex type cable or the armor cable. Flexible BX type cable is more expensive, but it does provide metallic insulation which is very strong over its wires and their insulation. Since it is flexible, it can be run in holes just as the Romex is run. Greenfield flexible armour conduit does not have wires and is usually not recommended.

Rigid conduit will probably have to be run in notches so that it will not project out and interfere with the placement of the gypsum wallboard. Notches must be covered with steel plates, especially if the conduit is aluminum or plastic. It takes skill to make the proper bends in the metallic conduit so it fits nicely. The plastic conduit can be bent with a little heat if needed. Also, the plastic has lots of fittings which can be used to get bits and pieces of the pipe routed to the places it needs to go.

## RUNNING WIRE AROUND A CORNER AND OVER BEAMS

Let us examine more of the wiring. In Fig. 6-17 we see the result of going around a corner in the drilling situation. Here the outside (really an inner side) wall of gypsum has been put up as you see in the background. A brick veneer will be put up outside against this inner outside wall so that, when finished, the home looks like a completely brick-walled house. With the large holes, the Romex

Fig. 6-16. Going around a corner in the framing means lots of holes. Holes must be centered at least 1½ inches from the edge of the studs so nails will not pierce the cable.

Fig. 6-17. Here is how Romex is "run" around a corner.

Fig. 6-18. Cables can be run over the ceiling beams in the crawl space or in the attic.

Fig. 6-19. A "tie-up" bit of wire holds a ceiling cable in its place. This is a multi-wire cable.

runs easily aound the corner and starts down one wall to its final box. Notice also that another Romex might be run alongside the one shown if two branches are necessary.

There are some cases where the wiring might be run over the beams supporting the roof or the second floor as shown in Fig. 6-18. These cables are not stapled down to the beams. They will be held in place by the starting and exit points or sometimes by a "tie-up" piece of cable as shown in Figs. 6-19 and 6-20.

You might have noticed in Figs. 6-18 and 6-19 that the cable runs over what appears to be sharp edges in the wood-fastening metallic squares. You probably shudder at the thought of pulling the cable past these edges and having the insulation scraped away and a short circuit developing. Be careful not to scrape the wires over these metal edges. I would use some staples and route the cable away from this dangerous point.

Routing wires over beams in an attic is very common as shown in Fig. 6-21. To the right you see a fixture for a pull chain type of light. These wires will not be for the air conditioner; they are 14/2 light cables. The air conditioning wires must be much larger, 8-2 at least. These wire cables are run through the attic on a point-to-point basis and may not be stapled down. Notice also that the wires have been run into the box, but no fixture is attached. The fixtures and receptacles will not be put into place until almost everything else has been done.

## USING PIPE FOR SUPPORT

In Fig. 6-22 we see how a corner in the framing is made to accept a single Romex cable. Notice how this cable has been wrapped around a pipe segment to the right. This kind of support is not that dangerous. The pipe surface is smooth and round. If the cable is not dragged across it, probably no damage to the cable will result. It would be easy to use some insulated staples and fasten

Fig. 6-20. A multi-wire cable is looped and tied to take up slack.

Fig. 6-20. A multi-wire cable is looped and tied to take up slack.

Fig. 6-22. A single Romex cable goes around a corner and a metal pipe.

the cable to the stud so that it never touches the pipe. But what we show here is common practice in home construction.

## STRANDED WIRES

Looking further into this home, we find the arrangement in Fig. 6-23 where a side wall intersects the back wall of what will be the kitchen. Here you can see the large (8/2) white cable which will

Fig. 6-23. The routing of a large cable on a run along a kitchen wall as it goes to the utility room.

Fig. 6-24. Going up a stud to a ceiling area for distribution routing. A switch may be below on the stud. Two lines may go to fixtures and one to the "home" location.

be used for a heavy duty outlet. This cable, incidentally, has *stranded* wires; otherwise, it would be too stiff to bend and run easily. Care must be taken when connecting all those strands to the receptacle, outlet or box. You do not want any loose or projecting strands left over after connecting to the screws. These projection strands might short out when you place the end device in the box or wherever. Notice that the wires run behind and away from the water pipe shown.

## THE LONG RUNS

For the long runs the professional always tries to get up overhead in the ceiling crawl space or attic space if this is possible. There he can make long straight runs and not have so much drilling through the wooden studs and beams to get the cables to the fuse box or circuit breaker location. In Fig. 6-24 we see how cables are run up a stud to a point where they can go into the crawl space. Then they are run, each in its separate way, to a light fixture, a switch or to the "home" location. In this installation there was no choice but to drill through the overhead beams to get the cables where they needed to go. When they had to come down, the top of the stud is again drilled and the cable comes down instead of going up.

We have another look at how you go through beams overhead with cables in Fig. 6-25. Notice that the holes are drilled at least 2 inches from the beam edge so nails holding the ceiling gypsum board will not penetrate into the wiring.

## CABLE COLORS

You might wonder why black cable is used sometimes and white cable other times. The answer is that the size wire in the

Fig. 6-25. Here's how cables are routed in ceiling joists.

Fig. 6-26. The ceiling cables must come down to the exit hole in the wall to get to the circuit breaker panel.

cable may be identified by the outside cable coloring. So it could be that the black cable has, say, 10/2 conductors while the white cable has 14/2 conductors. Of course, both cables also have ground wires along with their conductors. Also, you will find the individual wires in the cables coded black and white regardless of the color of the outside sheath.

Now in Fig. 6-26 we can see a ceiling run, its termination to a stud top and how the cables come down a wall stud. If you are wondering why the cable is twisted at the second beam, it is probably because the worker doesn't want the edge of that cable pressing against the sharp side of the hole in the wood as the cable is pulled along its designated path. The sharp side of the wood against the cable might have scraped away the insulation, just as a metal edge might do. It might also just be a sloppy job.

Some professionals might argue that cables can have any color with any size wire and what you get might be determined by which supplier you get it from. We acknowledge that possibility. Also, there might be a wiring reason for the different colored cables. If the installer wanted to make sure he knew which run went to a switch or light, he might use a different colored cable to show this location.

## GETTING CABLES THROUGH WOOD FRAMING

Next we examine the wall routing, going up overhead in Fig. 6-27. Here we see how some manipulation of the wooden support

Fig. 6-27. Sometimes the framing has to be cut to get the cables through, or specific places chosen to drill the cable run holes.

Fig. 6-28. The use of proper staples gives support and holds the cables in place. These staples will not damage the cable or its conductors.

structure had to be accomplished to get the three cables up overhead where they should be. Problems going through the wooden structure can often be very frustrating.

The use of staples to hold the cables in place is shown in Fig. 6-28. Never use nails pounded in on one side of the cable and hammered over the cable. Take care not to damage the cable.

## CEILING FIXTURE BOX

The fastening of a ceiling fixture box to a beam and its connection into a cable run amid many water pipes, is shown in Fig. 6-29.

Notice how the wires are tucked away inside the box. They won't be connected until after the ceiling gypsum board has been put in place and the room painted or wallpapered. You would suspect white and red wires here as this light undoubtedly will be turned off and on with a wall switch. That is the case as we see in Fig. 6-30A. A closeup of a ceiling fixture shows the red and white wires and the ground wire neatly coiled inside. A nail is used to support the cable, which is bad practice. In 6-30B we see another bad example of nail use. Fortunately, it was not hammered down into the cable.

Fig. 6-29. Wires go inside the light fixture box which is fastened to the ceiling joist with nails.

Fig. 6-30. Ceiling fixture wiring. (A) Closeup view of wiring inside the box. Note the improper use of a nail as a cable clamp. (B) A band installation uses nails.

When the box is to be set into the brick veneer outside wall, the box is simply strung on the wire which goes inside it as shown in Fig. 6-31. This box is for a back door light.

## WIRING OF VARIOUS LIGHTS

On the outside of the home, will be the recessed lighting in the overhangs such as shown in Fig. 6-32. This fixture has been

Fig. 6-31. A box to be installed in the brick veneer is hung on the cable at the back door. It is for the back door light.

219

Fig. 6-32. Box for an overhead light at the front entrance.

screwed to a support member with the two screws in the center. If you will notice the lip of the box, you will see four evenly spaced holes around the lip which can be used to fasten the box directly to a firm board like the plywood shown here. When finished, there will be a glass cover and light fixture attached to the wire which goes to the door switch. The roof overhang will be painted with a couple of coats of good water-resistant paint since this is a front doorway light location.

In some cases one might want to turn on two lights at the same time and from the same switch. You will have the line from the switch coming into the box. A second line (cable) goes out from the box as indicated in Fig. 6-33A. Notice that the connections for the extension cable are made inside the box. The white wire from the "hot" input cable to the switch is connected to the white wire of the cable to the light. This connection will be pushed into the back of the switch box and capped with an insulated electrical nut.

Large rooms such as dens or playrooms will have this multiple parallel lighting all operated from the single switch near the entrance doorway. Or a switch may be used that is connected to a "runner" cable (Fig. 6-33B). The white and black wires are hot in a runner cable. Be careful.

We have shown previously the recessed type of light which is found in bathrooms over the vanity or counter and in kitchen areas. An example of how this fixture will appear when it is wired is shown in Fig. 6-34. The fixture itself has a section of BX cable which is connected to the high temperature socket. These sockets

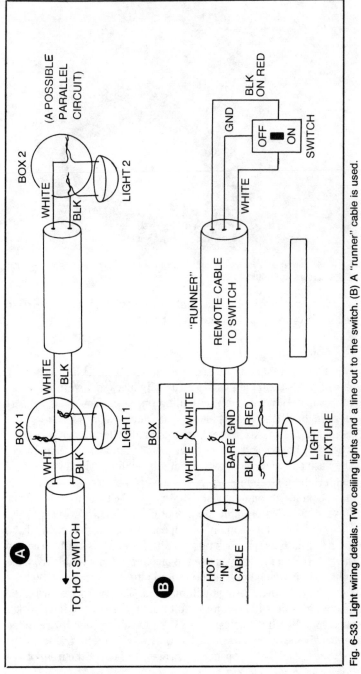

Fig. 6-33. Light wiring details. Two ceiling lights and a line out to the switch. (B) A "runner" cable is used.

221

Fig. 6-34. Wiring of a recessed light fixture using BX armored cable. Romex brings current to the fixture box.

are of ceramic or similiar material and are not of plastic. There are small vents to permit escape of hot air.

## BX CABLE

The BX cable, as you recall, is an armored cable which has a pair of wires contained inside it, usually black and white in color. You cannot remove them by pulling on them. You cannot insert them if, by chance or accident, you somehow do manage to get them out. A similar looking armor conduit called Greenfield cable is actually hollow and has no wires inside until you put them there. They are pulled through with a suitable fishline, which is a strip of metal ribbon. Greenfield comes in many diameters so many wires may be pulled through it, just as with rigid plastic or metal conduit. Some city codes prohibit the use of Greenfield, however.

The BX cable actually disappears to the upper right of the fixture (Fig. 6-34). It goes into a built-in metal box on the frame of the fixture. To the left you see the incoming Romex cable from the light switch which will also be routed into the box. Since this is a metal box, Romex fittings will be used. The two wires from the incoming cable will be connected to the two from the light socket and capped with electrical nuts. The incoming ground wire in the cable will be connected under a bolt to the metal frame of the fixture. Recall that an opaque glass plate will cover the front of this

222

fixture in a beautiful manner when the ceiling is put in place. The glass plate will be flush with the ceiling. It will be possible to remove it easily, however, to replace the light bulb. Notice the toothed edges of the hangers which permit locating the fixture anywhere between the beams.

## BELL TRANSFORMER

Sometimes we wonder where the bell transformer is located. Figure 6-35 shows one installation near an attic switch which has two functions. The first switch of the couple is used to turn on and off the attic light. The second switch turns on and off the electrical power to the relays which control the blower motors of the air conditioning-heating unit found in the attic. This latter requirement is so a person can work on that unit and turn off the power, while not having to run down stairs to the circuit breaker panel to turn the power off and on as he works. By having the air conditioner and heater blower on a separate circuit controlled by a switch there in the attic, the maintenance personnel can have light and power to use electric drills, if necessary, while cutting off the power to the air conditioning unit as they work on it.

Notice that the bell transformer is simply connected (bolted) to the cover plate of a metal box, and its 110 volt wires are connected inside the box. The lines for 110 volts then run up to the switches and, on the back side of the stud or 2×4, go on to the light

Fig. 6-35. Bell transformer and attic switches.

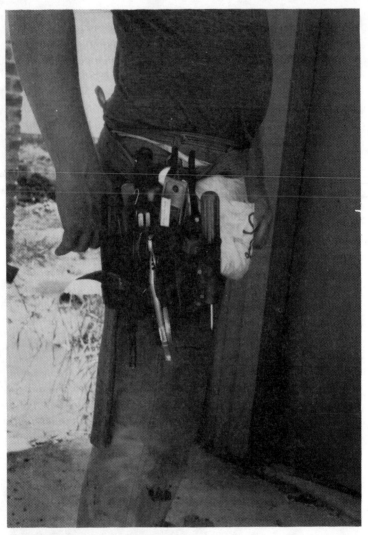

Fig. 6-36. This belt has room for many tools.

box and the air conditioning unit. The low voltage bell wires are simply stapled in place on various support members as they run to the switch at the front door and to the chimes. In Fig. 6-35 we show how these low voltage wires are connected to these devices.

## TOOLS OF THE TRADE

We have examined some wiring layouts and cable runs. Now we will look at the internal wiring in more detail. Many tools are needed as you can see in Fig. 6-36. Insulated handle pliers and screwdrivers, levels, an adjustable wrench, hammer, multi-clamp pliers, stapels, nails, a good knife or blade, a wire stripper and other special tools are utilized. In Fig. 6-37 we see a wire stripper which has notches for the various sizes of wire. When using this, you can cut into the insulation of a given size wire and not nick the wire itself, if the proper hole is used. A quick jerk of the wrist will then slide the insulation off the end of the wire you are working on. Notice the clamp projections just above the handle's insulation. These can be used to crimp various fasteners to the ends of the wire.

In Fig. 6-38 we see how one master electrician bent one of his screwdrivers so he could get a "crank" action by whirling his wrist around in order to seat bolts and screws. We were amazed at how well this worked and how quickly he seated the bolts on the air

Fig. 6-37. A wire stripper is a handy tool.

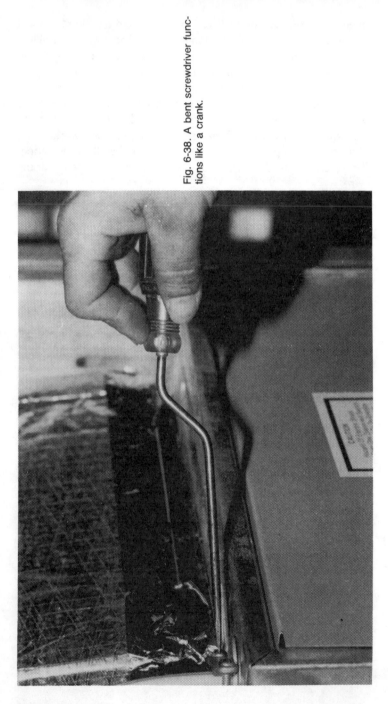

Fig. 6-38. A bent screwdriver functions like a crank.

conditioner he was working on. There are, no doubt, many other special tools devised by the men who put in the wiring, and they cannot do without them.

## TWISTING AND CAPPING WIRES

Look at Fig. 6-39 where you see five cables running down the stud into a switch box. Some, perhaps, not only seem to run down but actually are going back up and out somewhere else. Then look

Fig. 6-39. Five cables into one switch box are not uncommon.

Fig. 6-40. Many wire ends in a box.

at Fig. 6-40 where you see the mess of wires which will be coiled in this box until after the room paneling is put into place and painted or papered.

In Fig. 6-40 the bare wires are twisted together and capped with an electrical nut. Also, a number of black wires are stripped at the ends, connected together by twisting and capped. Back inside the box the white common wires are stripped at their ends, twisted together and capped. Notice that there are a number of "free" ends of black wires.

## SWITCHES

There are three switches needed here. A common "hot" line comes in and then the switches, one side of them, will be connected to this hot line. Then the other side of the switches will be connected to the black "run." In Fig. 6-41 the input hot cable is at (A) and it has three wires: one black, one white and one bare wire.

The hot black wire from the input goes to one side of each switch as shown. The other side of the switch is then connected to the black wire of the cable which goes to the light fixture. The white wire of the light cables are all connected together and capped, as are all bare wires of all cables. As you see the switches and lights are in series, but the series combination of each is then connected in parallel across the input hot lines.

You don't always have three light switches in a box. Make sure to use switches which have sufficient amperage to handle whatever load is used.

When switches are designated, all you have to do is pick out the incoming "hot" cable and then make the connections to switches as was shown in Fig. 6-41, whether it be one or four switches. Some professionals put a little notch or imprint in the insulation of the wires. For example on *travelers* (connections *between* switches at ends of hallways), they put three little indents.

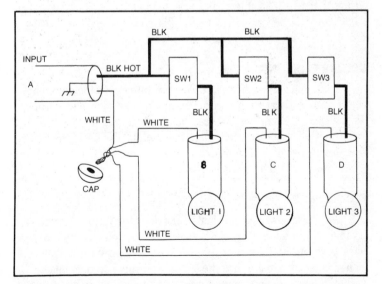

Fig. 6-41. A wiring diagram for three switches and three lights. A, B, C and D are Romex cables.

Fig. 6-42. Indents are used on wire insulation. The end of a "hot" wire is shown along with some pass-through wiring.

Some electricians always strip the end of the hot black wire coming into a box. But they don't strip the ends of the other wires, except the white ones which are connected together and then capped.

How do you know which pair of wires goes to what appliance or light? Size 12/2 cable goes to the dishwasher box. A size 14/2 cable goes to the garbage disposal. Heavier or larger size cables are used for electric dryers water heaters and the like.

## GROUND FAULT INDICATOR

A ground fault indicator (GFI) goes into every location where there may be water: bathrooms, outside porches and garages. There can be a number of receptacles which might be controlled from the GFI. Lights or lighting circuits might also be controlled by this device. A little red button snaps out from its flush position when the device has been tripped.

## CEILING FANS

Ceiling fans are easy to install and don't use any more electricity than a light bulb. They can help reduce your electricity bill. You wire them into a light fixture, just like the light fixture is wired. You may have to make sure the box is well supported on beams, or you might have to put some bracing in the ceiling to support the weight of the fan. They use about 1.5 amperes of current. Some

230

Fig. 6-43. A pass-through wiring situation for a bedroom receptacle. Notice the "P" designation.

types are sealed and don't require any maintenance. Others need oiling once in a while.

## PASS-THROUGH WIRING

A glance at Fig. 6-42 shows the end of a hot black wire protruding from the box. The end of this wire has been scraped and the insulation taken off so that its identity is unmistakable. We also see some "pass-through" wiring for a receptacle past the highest one. To the right is another receptacle on a completely different branch circuit.

## NEW SMOKE DETECTOR ORDINANCES

It's happened in Houston and will probably happen elsewhere. New ordinances are requiring that all new homes have a smoke detector in each bedroom in addition to having one at stairways and in kitchens. Use of battery units makes the smoke detectors independent of electrical input from mains which might, somehow, cause electric current to be cut off. You must check these batteries at least once a year. If you cannot check them, simply plan on replacing them each year at a specific time so you won't forget. If you feel you want to have some units operating from the electric current lines, be sure you do have some "mix" of types so that others will be battery-operated. This gives you the best protection. Check your smoke detectors at least once every month for correct operation.

## BEDROOM AND LIVING ROOM WIRING

In Fig. 6-43 we see how a box for a receptacle is located on a stud in a bedroom wall. The line going to it is duplicated by a return line (cable) which goes on to another box in another wall. Notice how the cables are stapled in place and how they run through center holes in the stud.

Since other boxes will be connected in parallel with this one, the return cable is used. Sometimes, when the boxes are located on walls beyond where the box is located, the second cable simply goes on—toward the left in this case—to the next box.

To see how the cables go into the box, we look at Fig. 6-44 and find that there are no fittings used. Fittings are not generally used if the cables go into a plastic box. If the box is metal, you should get Romex fittings.

Just because some boxes are "back-to-back," they do not necessarily connect to the same branch circuit. Different rooms

Fig. 6-44. Romex going into a plastic box uses no fittings.

have different circuits to balance the loading and prevent a total blackout of a given room if the branch circuit somehow is cut off from its electrical supply. A case in point can be seen by studying Fig. 6-45.

On the left of Fig. 6-45 are two lines coming in, and one goes to the left-hand box. The box to the right gets its current from a

Fig. 6-45. The wiring of back-to-back boxes on adjacent studs. Several branch circuits are involved.

Fig. 6-46. The height of the cable run is shown.

cable coming in from the right. The right-hand box also is connected in parallel with other boxes, so there is a second cable coming out of this box and going to the left. The box on the left is to be operated from a light switch near the doorway, since it faces the living room. A second line has to be run to the switch for that purpose.

There will be lots of wire connections behind the receptacle of the box at the left. Notice that the wire cables are run higher than the level of the boxes. This is practically always true. Cables are

Fig. 6-47. Multiple cables for two boxes located close together is not uncommon.

235

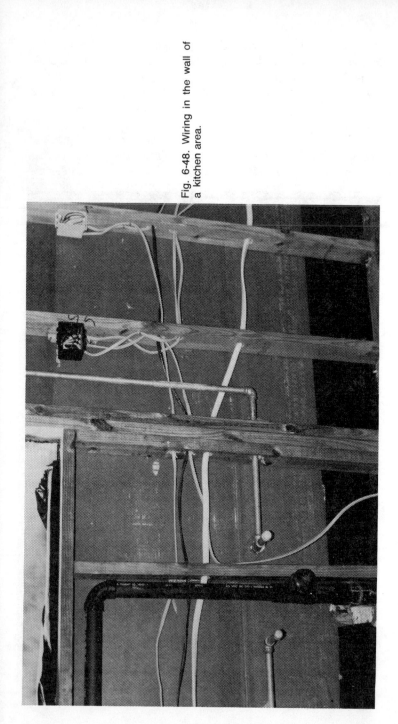

Fig. 6-48. Wiring in the wall of a kitchen area.

not run along the floor level or at the box levels. In Fig. 6-46 you get a better idea of how the cables are run for boxes and at what level they are installed.

To show the complexity that sometimes exists over a given run, look at Fig. 6-47. Here we have a switch box above a receptacle box. Lines coming into and out of the lower box show it is in parallel with other boxes. The number of cables into the upper box indicate it is probably a three or four-way switch arrangement. This pair of boxes can be found in a living room near a doorway. The pipe going down the wall is a water pipe going to a nearby bathroom.

## KITCHEN AND UTILITY ROOM WIRING

Now we examine a kitchen area, especially the place where the sink will be located. See Fig. 6-48. Notice the drain pipe and

Fig. 6-49. An electrician at work in a hot location.

237

Fig. 6-50. Cables going into and out of an air conditioning-heating unit.

the pipes for the hot and cold water outlets. You will see a length of cable going to the floor from near the right-hand water pipe. This will be for the garbage disposal. You see the wide white cable which is larger than the others. This is an 8/2 pair which will connect to the next room for the electric dryer and washer.

At the right of the window will be a countertop. Above this will be the switch for a light and the garbage disposal. This is the black box. At the right of that is the necessary countertop outlet which will be on a heavy duty circuit (20 amperes). It is in parallel with another which will be on a counter to the right that is not visible. The lines running to the left are for the utility room, which is located to the left of the kitchen area.

As we examined this building, we were fortunate enough to meet a master electrician at work, installing the blower of an air conditioner-heater (Fig. 6-49). In Fig. 6-50 is a closeup of his wiring efforts. In Fig. 6-50 you can see the incoming cables and the switch and control cables which had to be connected to the wiring. Notice the use of electrical nuts on the ends of the connections.

In Fig. 6-51 you can see how the ends of the wires are twisted together before the "nut" is screwed in place. These plastic caps or nuts are very strong. They are somewhat flexible, if you get the right kind, and cannot be broken by being compressed in the jaws of pliers. The metal spring which clamps the wires together tightly when the cap is in place is built into the cap and probably will not have to be replaced. If corrosion occurs which would loosen the electrical connection, then get a replacement.

## CABLE TELEVISION CONNECTIONS

We came across television cable antenna lines all bunched together on the outside of an apartment building, ready to be run to a cable television distribution system or an outside antenna distribution system. Running these cables is important in electrical wiring. You must not bend them sharply, clamp them tightly with staples, or make long complete electrical "runs" with them. They need straight short runs as far away from other wiring as possible. The whole system needs a common ground for the grounding braid in the cable. The wires shown come from all apartments of a particular complex to the central distribution point where an electronics expert will make the final connections. Examine Fig. 6-52.

## INSIDE WIRED ROOMS WHEN WALLS ARE FINISHED

When you consider that mess of wiring which goes into the walls of a home or apartment, you begin to wonder what it will look like when the walls are finally put into place. We will examine some illustrations of that state of development. We pay credit and due respect at this time to those highly trained professionals who put that wallboard in place. How they are able to take a 4x8 foot sheet

Fig. 6-51. Ends of the wires are twisted together before the nut is screwed in place.

Fig. 6-52. These antenna lines will go to a central antenna distribution system.

of gypsum or sheetrock and determine exactly where the socket, switch and receptacle cutouts are to be fit will forever remain a mystery to me. Examine Fig. 6-53.

In Fig. 6-54 you see one switch wired in place. Notice the fit of the wallboard around the electrical boxes. For a closer look at the fit, see Fig. 6-55. There we see the switchbox and one outlet. Pipes, cables, connections and fittings are all there behind that gypsum wallboard.

In Fig. 6-56 you see one side of the living room and the two outlets, which, you will remember, cannot be over 12 linear feet apart. In a living room, a receptacle can have one of its two outlets connected to a switch. You can use lighting from a lamp instead of from an overhead light, if desired and turn it on and off from the doorway.

The kitchen, breakfast room, bedrooms and baths will all be finished in the same manner. In the kitchen area the ovens will be installed and connected into their concealed boxes. The range will also be connected and installed in its countertop. The counters with formica will be put into place and the whole building will be ready for the finishing touches. See Figs. 6-57 and 6-58.

## MORE DETAILS ON CIRCUIT BREAKER AND METER PANELS

We have been through the house and examined its wiring in fine detail. We have mentioned many times that all circuits will "run" from their terminal devices to a "home" location. There they

Fig. 6-53. The wallboard in on place over the switch and receptacle outlets.

will be connected to the fuse panel or circuit breaker panel. We need now to examine this aspect of home wiring. As we start, let us examine the interior of the home again to see all those cables coming together at one location on one wall. Then they exit the building and go outside. There are many of them and to determine which cable comes from, or goes to, where might be an unsolved mystery if we didn't have some help. Examine Fig. 6-59.

Notice that at the point of exit there is the familiar designation (S,S) for a double switch to be installed on the vertical stud. Just below the exit point and to the right of the stud you will see the box.

Fig. 6-54. Final wiring of the switch after the wallboard is completely installed and painting is done.

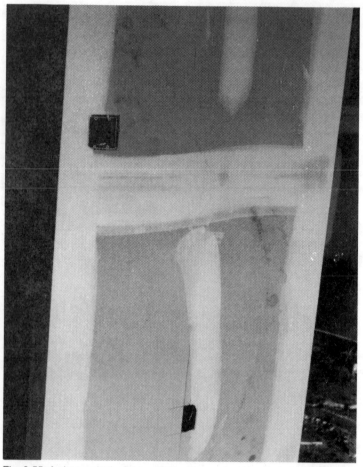

Fig. 6-55. A closeup look at the switch and receptacle boxes in the unfinished wallboard would show neat work.

Look at Fig. 6-60. To the right of the exit cables is the underground service from the electric company's tranformer with its large 100 ampere capacity lines. The pipe for the serivce is, incidentally, a PVC plastic type. As you know, circuit breakers are in two lines in a panel. In smaller dwellings such as apartments, they just have one line of breakers. As you see in Fig. 6-61 there is a master breaker at the top. When this is moved to the off position, it cuts off all electricity to the other breakers and to all the house circuits.

On the left you see, at the top, a pair of breakers. This system is used in many cases for high amperage lines such as those going to

Fig. 6-56. The living room outlets are properly spaced 12 feet apart.

the air conditioning-heating units or washers. The double unit may be replaced by larger single units. You can identify them by their physical size usually if they are not labeled in writing.

Most of the other breakers will be single units of the 15 to 20 ampere size as shown (Fig. 6-61). They are clearly labeled off and on in the two positions which their bars can have. Below the

Fig. 6-57. Connecting a light fixture in the finished bedroom.

243

Fig. 6-58. Installation of the light fixture completes the job.

indicated breakers are some metal coverings called "push-outs" which permit removal with a screwdriver tip for the addition of more circuit breakers when desired. When a breaker trips, it simply moves its bar from the on position to the off position. Reset by pushing it back to the on position. If it will not hold in the on position, there is trouble in the branch circuit or possibly the circuit breaker. Call a master electrician to check the circuit for trouble.

There is no wiring put in yet inside the small circuit breaker panel shown in Fig. 6-62. The large round inlet at the upper left is where all the cables will enter. Notice that there are actually no breakers installed yet. There is no master circuit breaker in this box, which is in an apartment, as the master switch is outside in a box of its own.

### Metal Nubs

Look in the center of Fig. 6-62 where you see small metal nubs which are bent as though to receive and hold an object. One end of the circuit breaker will fit here. The center of the box has the hot bars. The one on the left of the nubs is uninsulated. The center bar is separated from a fitting, near its center, by plastic insulation. To the right are four more "nubs." This panel will accommodate at least eight circuit breakers. At the bottom you see a bar with many screw heads. This is the ground or common bar to which all white wires will be connected. The black hot wires of cables will each be connected to a terminal on the circuit breakers. Let's examine the installation of a circuit breaker so we see how the nubs and bar connections are made. See Fig. 6-63 and Table 6-1.

## Connections to Breakers

You *must* use a circuit breaker designed for the specific panel which you may have installed in your dwelling. Not just any circuit

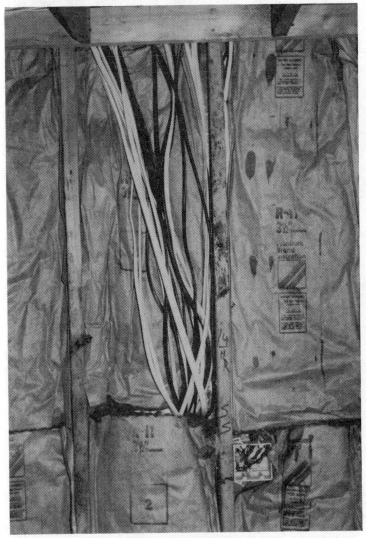

Fig. 6-59. The ''home'' location of all branch circuits.

Fig. 6-60. Circuit breaker panel wires and cables are shown along with the location of the electric meter socket. The underground riser of the service feed is also shown to the right.

breaker will fit into any panel. Notice in Table 6-1 the various sizes and types of breakers which are available. Although in Table 6-1 it is indicated that aluminum wire (AL) may be used, we do not recommend this type of wiring. Even so, if you have a home wired with aluminum wire and it comes to the circuit breaker panel, you must have a breaker which is designed for this type wire. The breaker will have the CO/ALR label. Even though other breakers may work, don't use them with aluminum wire unless that CO/ALR label is on the package or device itself.

If you have a cable with 14/2 wires in it, you connect it to a 15 ampere breaker. If the cable has 12/2 wires in it, which goes to a toaster or some appliances, you connect it to a 20 ampere breaker. A white 8/2 cable for an air conditioner is connected to a 60 ampere breaker. It takes that many amperes to start the motor. The motor will run on about 20 to 30 amperes after it starts. But you have to use a big capacity breaker so it won't trip when the motor starts. A 10/2 cable with ground is connected to a 30 ampere breaker. These ground wires (bare ones) are connected together and then attached to a ground rod. Use a ground rod which is 8 feet long and drive it 6 feet down. You can also connect to a water pipe system for a second ground. Make sure the water pipes are grounded in the home also. Sometimes they come in from the mains in plastic, and then have

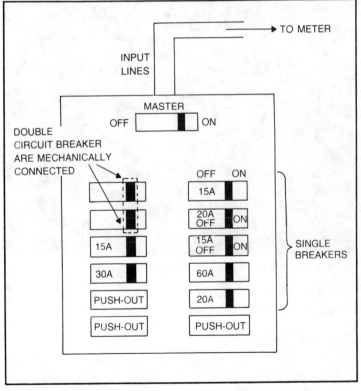

Fig. 6-61. A diagram of the layout of the circuit breakers in a panel.

Fig. 6-62. A view inside an apartment circuit breaker panel before it is wired.

iron pipe inside or copper pipe. Connect them to the ground rod. Examine Fig. 6-64 for an example of a ground rod.

## Panel Arrangement

How are circuits and breakers arranged so that the home gets an even distribution from the split input service? The split input has two large wires and a smaller, common, third wire which is

248

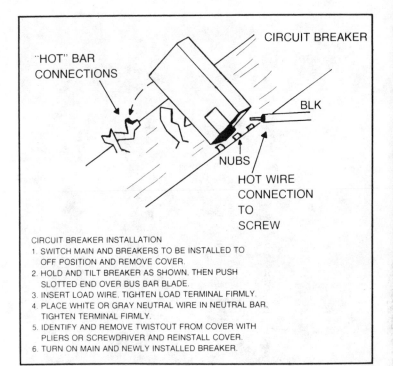

CIRCUIT BREAKER

"HOT" BAR
CONNECTIONS

BLK

NUBS

HOT WIRE
CONNECTION
TO
SCREW

CIRCUIT BREAKER INSTALLATION
1. SWITCH MAIN AND BREAKERS TO BE INSTALLED TO
   OFF POSITION AND REMOVE COVER.
2. HOLD AND TILT BREAKER AS SHOWN. THEN PUSH
   SLOTTED END OVER BUS BAR BLADE.
3. INSERT LOAD WIRE. TIGHTEN LOAD TERMINAL FIRMLY.
4. PLACE WHITE OR GRAY NEUTRAL WIRE IN NEUTRAL BAR.
   TIGHTEN TERMINAL FIRMLY.
5. IDENTIFY AND REMOVE TWISTOUT FROM COVER WITH
   PLIERS OR SCREWDRIVER AND REINSTALL COVER.
6. TURN ON MAIN AND NEWLY INSTALLED BREAKER.

Fig. 6-63. Installation of a circuit breaker.

Table 6-1. Information on Circuit Breakers.

| SEARS NO. | RATING | | +WIRE SIZE | | TYPE |
| | AMPS. | VOLTS | CU | *AL | |
| --- | --- | --- | --- | --- | --- |
| $^{34}$5330 | 15 | 120 | 14 | 12 | **Std. single pole |
| $^{34}$5360 | 15 | 120 | 14 | 12 | **Two single poles |
| $^{34}$5331 | 20 | 120 | 12 | 10 | **Std. single pole |
| $^{34}$5361 | 20 | 120 | 12 | 10 | **Two single poles |
| $^{34}$5332 | 20 | 120/240 | 12 | 10 | #Std. double pole |
| $^{34}$5333 | 30 | 120/240 | 10 | 8 | #Std. double pole |
| $^{34}$5365 | 30 | 120/240 | 10 | 8 | Quad-two dble poles |
| $^{34}$5334 | 40 | 120/240 | 8 | 6 | #Std. double pole |
| $^{34}$5335 | 50 | 120/240 | 6 | 4 | #Std. double pole |
| $^{34}$5364 | 60 | 120/240 | 4 | 2 | #Std. double pole |
| $^{34}$5210 | 100 | 120/240 | 1 | 1/0 | #Std. double pole |

  + Minimum recommended sizes for 60° C insulation, Table 310-12 & 14,
    1974 National Electrical Code.
  * Also applies to Copper-clad aluminum wire.
 ** 15 Amp. and 20 Amp., 120 volt breakers are widely used in lighting and small
    appliance circuits.
  * Double pole breakers are commonly used in high capacity circuits like ranges,
    dryers and waterheaters, etc.

Fig. 6-64. A ground rod for a home.

connected to the ground system. The voltage is 220 between the two large wires and 110 from each large wire to ground. The circuit breaker panel is arranged so that if you go down the row, one breaker after another, one breaker connects to one main, the next breaker connects to the second main and so on. See Fig. 6-65.

The way the mains connect to the two bars, and thus to the vertical stubs which make connection to the breaker parts, is shown in Fig. 6-66. The common, white wire for the mains goes to the left and down to the bottom of the box where the ground bar is located. This connects to all white common wires. A view of the ground bar at the bottom, and above it the nubs on the left and right

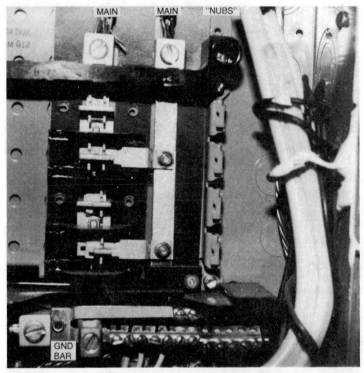

Fig. 6-65. Closeup of a circuit breaker panel. The two input "mains" are split in the row on the left.

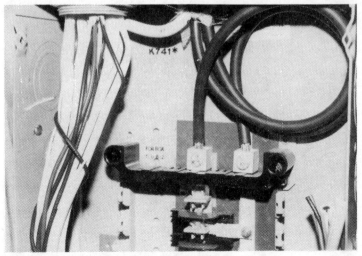

Fig. 6-66. Connecting the black mains at the top to the bars.

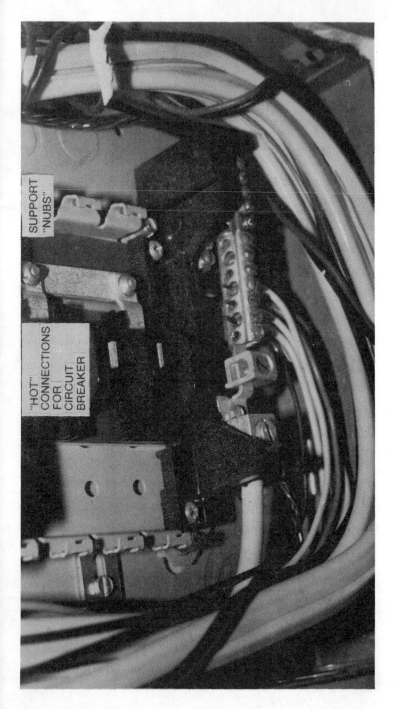

SUPPORT "NUBS"

"HOT" CONNECTIONS FOR CIRCUIT BREAKER

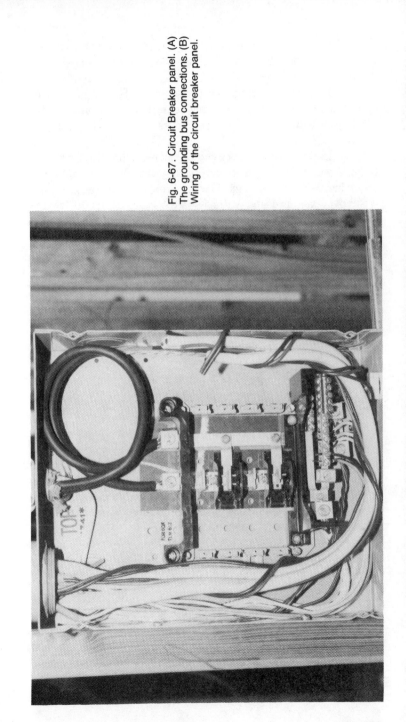

Fig. 6-67. Circuit Breaker panel. (A) The grounding bus connections. (B) Wiring of the circuit breaker panel.

253

side of the hot center connections for the breakers, is shown in Fig. 6-67A. Notice how the branch circuit white wires are pushed up through holes in the ground bar. Then screws are tightened down on them to make firm electrical connections. The large service, white common wire can be seen connecting to the ground bus on its left end. In Fig. 6-67B we see the wires to be connected to the circuit breakers.

We do not see the main circuit breaker which is used in homes and into which the mains go before they travel to the individual circuit breaker bars. That main circuit breaker is located at the top of the panel and probably will have a total capacity of 100 to 200 amperes. All electric current for the home must go through it. The master switch is not shown. There is another master switch, located elsewhere in this building, which fuses the main lines before they are run to the small panel. In Fig. 6-67B we see the panel installed, except for the actual circuit breakers themselves. Notice that all black wires of the branch circuit cables are not connected. You can see some of these wires on the right, wrapped around the white cables (Fig. 6-67B). These will connect directly to the circuit breakers on the circuit breaker-terminal connection (Figs. 6-68A and 6-68B). The hot wire goes to the circuit breaker terminal shown. The panel fastener slot fits the panel nub.

## Replacing a Circuit Breaker

If we have to replace a circuit breaker, we must use the correct size unit. We must find and open the master switch to disconnect all power before we go inside the panel for any reason. On most home circuit breaker panels, it is customary to remove the meter from its socket. This cuts off all power to the home until it is replaced. Of course, if a master electrician does this job, he has a seal to use for reinstallation.

## Inside the Circuit Breaker

The circuit breaker is sealed and cannot be opened, but we opened one to show you what is inside it (Fig. 6-68B). At the top is the bar switch lever with which we turn it on and off. Down in the lower left we can see the terminal which is connected to the black wire of a circuit. Just to the right of that we see two points which come together to make a connection and fly apart when the breaker is tripped. We can also see the magnet, its coil and the mechanical mechanism which trips the breaker when an overload of current

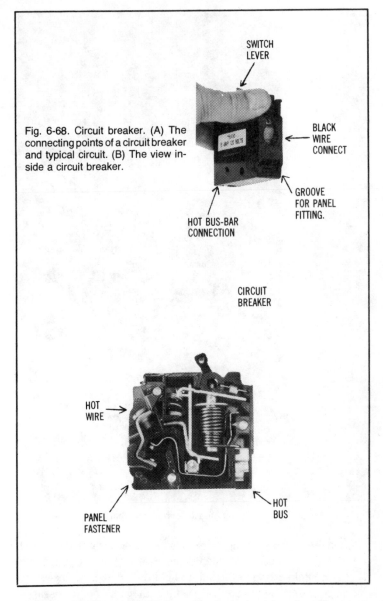

Fig. 6-68. Circuit breaker. (A) The connecting points of a circuit breaker and typical circuit. (B) The view inside a circuit breaker.

SWITCH LEVER

BLACK WIRE CONNECT

GROOVE FOR PANEL FITTING.

HOT BUS-BAR CONNECTION

CIRCUIT BREAKER

HOT WIRE

PANEL FASTENER

HOT BUS

passes through that coil. This magnet is designed so that all current controlled by the breaker passes through it. It will not attract the trip metal lever unless the current exceeds its rated value. The springs shown control the mechanical functions of the trip lever at the bottom.

Fig. 6-69. Installation of a circuit breaker panel in a "rough-in."

On the far right, at the bottom, is the connecting clamp for the bar (or bus). At the far lower left is the plastic cutout that slips into the horizontal section of the nub to hold the breaker on that side. Although the circuit breakers are sealed, they may not necessarily be airtight. In time, their small springs may change value. One tends to forget about them over the years if you do not have trouble with the electrical system.

The installation of a small electrical panel such as we have shown is given more consideration in Fig. 6-69. It is simply fastened to a stud, just like any box but more strongly, using screws. This is an eight circuit panel which, while small, is large enough for small homes or apartments.

### Circuit Breaker Panel Overload

The twin service input and the way the panel itself divides up the load make it very unlikely that either leg of the main service input will be overloaded. We have said that the input mains would be probably 100 amperes each. The total load for the lights if everything is on might run as high as 8,000 watts. That's about 80 amperes. One hundred amperes per leg of the mains is plenty of current.

### A Home Circuit Breaker Panel Installation

Let's examine Fig. 6-70. We see the wires of all the branch circuits of the home alongside the underground service input (PVC plastic pipe) emerging from the brick veneer wall, which is placed outside the wooden framing. All that is needed here is the meter socket, meter and circuit breaker panel. All the wires of the service input are black. How then do we determine which one is the common wire? That is easy if you know the secret. When you have three wires like this, there will be one which is smaller than the

Fig. 6-70. Meter and circuit breaker panel location for a new home. The service riser is to the left.

257

Fig. 6-71. The meter socket is now installed.

others. That one is the common wire. The way the job looked when the panel and meter socket were installed is shown in Fig. 6-71.

A closeup of the circuit breaker panel in Fig. 6-72 with the mains designated at the upper right and the air conditioner units on the upper left. Below each of these are the appliance and lights breakers. The main breaker designation here is the master switch for all circuits.

If you live in a humid, damp area where moisture may get inside and corrode the elements, have your circuit breakers checked at least every year. Under really bad moisture conditions, it might even pay you to replace them every year or so. The cost might be small compared to what might happen if they fail to operate as they should.

## TRANSFORMER

Let us return to Fig. 3-1, where the installation of the electric company's service is being accomplished. The method used is an

Fig. 6-72. Designation of circuits inside a circuit breaker panel.

Fig. 6-73. The underground feed transformer is now wired. No connecting lines are visible.

underground feed system from a transformer. You have responsibility for all lines to the transformer. The transformer itself is the responsibility of the electric company.

In Fig. 6-73 we see how the site has changed with the addition of a backyard fence and the underground cable system all connected and wired to the transformer. There are no visible connections anywhere. Alongside is that telephone company's terminal which connects to many telephone lines in this area. The ground, which has been disrupted, will settle. Often this causes a shifting of the terminal. Notice how it leans in this illustration. Sometimes the ground settling will cause the transformer concrete slab to sink or to tip at an angle. In these cases you probably would want to call the electric company and see about getting an additional slab or an alteration.

## SUMMARY

So we come to the end of this important chapter. You should have a good idea now of how wiring is accomplished in a home, and you will understand why it is not an easy task to rewrite a home. After careful study of this chapter, you will probably conclude that you can do the wiring yourself. That may be a valid conclusion. You may live in or near a city where it is necessary to have a master electrician do the work in order to get a permit and acceptance by the city electrical inspector. Be sure you have the electrician either do the work or possibly supervise your work if you do it yourself.

# Electrical Lighting

7

We cannot do without light. We must have it day or night. It must be of the proper level and intensity. Hopefully, it will be right with regard to color. We need to take a detailed look at the effects of lighting. Lighting fixtures and arrangements are some of the things we can change in our homes, perhaps to make them more livable. We may want to change fixtures in one room to create mood, in another to gain efficiency, and in another to permit work of some detailed nature or deep, hard study without eye fatigue.

It turns out that it is not always best just to get a bigger bulb if a room seems somewhat dark or your eyes start hurting. Perhaps the type and kind of bulb used, its placement and adjustment may be at fault. Among the fixtures we want to examine are track lights, fluorescent lights and even the outside lighting. We will try to explain the reasoning behind what is used and why.

## DEFINITION OF LIGHT

Light is defined as a radiation that can affect the eye. It is a vibration among many other vibrations, and its place in the *spectrum* is shown in Fig. 7-1, covering the vibrational frequency of visible light. There are other vibrations which we know exist but cannot see. Those vibrations which affect the *rods* and *cones* in our eyes send electrical impulses to our brains which, in turn, define the lighting, shading and coloring of what we see.

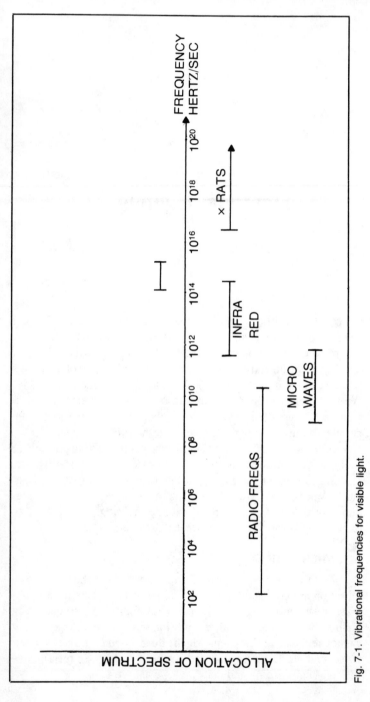

Fig. 7-1. Vibrational frequencies for visible light.

264

Notice how small the spectrum of what we can see is when compared to the whole of the vibrational spectrum. Within this small band are those frequencies of vibration which cause our rods and cones to inform our brains that what we are seeing is red, green blue or whatever.

## STANDARD VISION

Every person is different. We hear, speak and see differently. There is a graph which shows how an assumed "standard" person's eyes are affected by the various vibrations which make up the visible light spectrum. That chart is shown in Fig. 7-2.

The "standard" eye perceives light best in the yellow-green vibrational region. Almost 100 percent of this kind of vibration is "seen." This means that the eye will perceive smaller intensities of

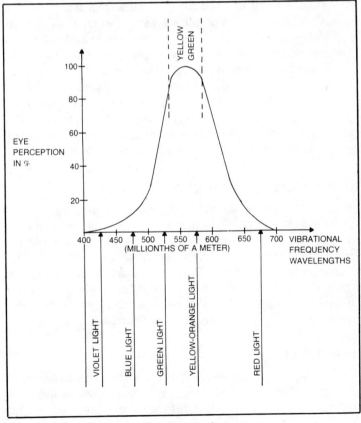

Fig. 7-2. Graph of how the eye perceives colors.

this kind of light then it will any other "color" of light. We use quotes on color because we want to replace the technical term vibrational frequency with the word color. Remember that the color is determined by how fast the vibrations take place. In summary, then, we see yellow-green light easier than anything else if we have standard vision. To put it another way, we can detect these colors or this color range easier than we can any other colors if we have standard vision.

But we do not all have this standard vision. We are all affected somewhat by environmental conditions and by various other factors too numerous to mention that cause our eye sensitivity to vary from the graph in Fig. 7-2. A person might have a complete loss of perception, say, in the red, blue or possible even in the green-yellow region. In this case the person may be considered color blind, or partially so.

How does all this relate to our subject in this book? You guessed it. It means that people will have different ideas about the type of lighting which must surround them. What is ideal for some persons may not be ideal for others. This can be true not only of colors but also of intensity. Imagine that you visit a home which has been decorated with colored lights. You might find the lighting garnish, irritating, too brilliant or direct or too soft and dim. But it is suitable for the eyes of the people who live there. Since you are aware that there will be differences in light perception by people you will tolerate the situation and be graceful about it.

## LIGHTING NEEDS OF THE ELDERLY

Let us examine the needs of some older persons. You will probably agree the lighting situations have to be different for these people. Did you know that the average 60-year-old person needs seven times as much light as the average 20-year-old person? It's true according to the American Optometric Assocation. Changes in the eye occur due to age. Older persons find more comfort in the light if it is yellowish (incandescent or directly lighted bulbs) rather than the blue light which comes from fluorescent lighting.

## ULTRAVIOLET RAYS

We don't often think about the colors which may be generated by light bulbs or produced through a secondary illumination when the light from a given bulb reflects from some colored surfaces. But they are there.

Note, too, that lighting which produces ultraviolet rays can be dangerous. You must be careful when you use high wattage or high intensity lamps of the arc or "special" illumination types that they do not give off any dangerous rays. It is true that the irritation from ultraviolet radiation may not last very long, unless one is constantly under and in its influence. It doesn't hurt to have this high intensity lighting checked.

## COLOR PRODUCTION WITH LIGHT

We need to know more about the color production which results when lights of various colors are mixed, or a light of a given color is reflected from a differently colored surface or object. Our mood and sense of well being are subtly influenced by the colors we perceive around us. Just think for a moment about a stroll in a forest

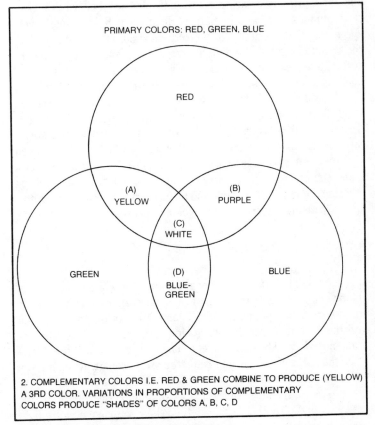

PRIMARY COLORS: RED, GREEN, BLUE

RED

(A) YELLOW

(B) PURPLE

(C) WHITE

GREEN

(D) BLUE-GREEN

BLUE

2. COMPLEMENTARY COLORS I.E. RED & GREEN COMBINE TO PRODUCE (YELLOW) A 3RD COLOR. VARIATIONS IN PROPORTIONS OF COMPLEMENTARY COLORS PRODUCE "SHADES" OF COLORS A, B, C, D

Fig. 7-3. Primary colors and blending effects.

area or park where countless shades of green are infiltrated with the coloring of flowers, grasses and skies. Think of the mood which comes over you in such an environment. Lighting in a room creates mood and atmosphere.

## Primary and Complementary Colors

Examine Fig. 7-3. Shown are the primary colors and the complementary colors which are generated or produced when the primary colors are mixed. You know that your television produces only the primary colors in the tube itself. It is the mixture in magnitude and kind of colors which gives you the various other colors which your eye perceives on the screen. We show only the complementary colors of first order magnitude.

You will get all the shading possible of these colors if the amount of primary coloring is varied throughout its entire range— red from maximum to minimum while green is held at some average level. Then vary the green from maximum to minimum while red is held at some constant average level. Naturally, if you are serving dinner and have some nice juicy steaks on the table, a red coloring enhances the appearance of that food on the table.

How does our light problem fit into this situation? If you use a lamp of a given color and filters (colored glass) in front of that light, then you will get various shades of coloring, just as the stage, theatre and movie houses do with their spotlights. As you can imagine, using a proper color on an outside lighting system of a home can enhance the coloring of, say, a red brick wall. Getting coloring effects on shrubbery and trees might be better done with a spotlight using a filter which gives a soft green glow.

## Reflection and Absorption of Colors

When light falls on a wall or other object in the home, some of that light will be reflected away and some of it will be absorbed by the object or "target" of the light rays. The color which you see in the object will depend upon the composition of the incident light and the extent to which some colors are reflected, absorbed or transmitted away from the object. The object itself will then influence the coloring of the light reflected away from it, while its own colors will be modified, enhanced or changed according to the composition (in colors) of the light falling upon the object and the intensity and direction of that light. When you have a white light and you place a sheet of red glass in front of it, the red glass will absorb almost all colors except red and orange. A sheet of blue glass will absorb many colors except those in the primary region.

If you have some object which you are focusing a white light toward and that object reflects all the wavelengths of light falling upon it, it will appear to be white in color. If it reflects none of the light incident upon it, it will appear black in color. Blue and yellow subject matter may reflect a sort of green color. The amount and coloring of light on a painting, picture or room which has many colors of tapestry, rugs and furniture will produce a big effect on what the person sees as he looks at the room. Remember, however, that what they see may not be what you see or anyone else sees as our eyes may all have some slight differences.

We have said that the optic nerve spreads over the retina of the eye into the rods and cones which exist there. The cones give us sight in daylight, and the rods are those elements which give us sight at night (twilight vision). The cones are the elements responsible for our color perception capability. They discern the mixture of light colorations and send those impulses to the brain. When a person has some color blindness, then the cones which are sensitive to a primary color may be weak or missing.

In some areas of the home such as the kitchen, shop, utility room, etc., you will want strong, bright white lights. In rooms where mood becomes important, you may want to vary the coloring in the room through change of lighting or use of filters on spotlights which highlight walls, tables or areas.

## EFFECTS OF BAD LIGHTING

Now let's find out about what is bad about lighting. It has been written that seeing is done by the eyes, nerves, muscles and the brain. Prolonged work in too little light or in a *glare* can be as tiring as working all day at some physical, tiring task. You can come down with eyestrain and nervous fatique. We bring up the concept of glare which has been defined as direct or reflected light that is visually disturbing and may interfere with seeing. Recall looking at a high gloss printed page in a magazine when the light was placed wrongly or you had the glare of the light reflected right into your eyes. Do you remember sitting at a highly polished table when some light source produced a reflection right into your eyes? You can probably think of other examples of bad glare.

## FACTORS GOVERNING SIGHT

How and what do we see? It has been written that there are four factors which govern our ability to see something. First is the

Table 7-1. Percent of Light Reflected From Various Colored Surfaces.

| Color | Approximate percent reflection |
|---|---|
| **Whites:** | |
| Dull or flat white | 75-90 |
| **Light tints:** | |
| Cream or eggshell | 79 |
| Ivory | 75 |
| Pale pink and pale yellow | 75-80 |
| Light green, light blue, light orchid | 70-75 |
| Soft pink and light peach | 69 |
| Light beige or pale gray | 70 |
| **Medium tones:** | |
| Apricot | 56-62 |
| Pink | 64 |
| Tan, yellow-gold | 55 |
| **Light grays** | 35-50 |
| Medium truquoise | 44 |
| Medium light blue | 42 |
| Yellow-green | 45 |
| Old gold and pumpkin | 34 |
| Rose | 29 |
| **Deep tones:** | |
| Cocoa brown and mauve | 24 |
| Medium green and medium blue | 21 |
| Medium gray | 20 |
| **Unsuitably dark colors:** | |
| Dark brown and dark gray | 10-15 |
| Olive green | 12 |
| Dark blue, blue-green | 5-10 |
| Forest green | 7 |
| **Natural wood tones:** | |
| Birch and beech | 35-50 |
| Light maple | 25-35 |
| Light oak | 25-35 |
| Dark oak and cherry | 10-15 |
| Black walnut and mahogany | 5-15 |

*size* of the object of focus. The larger the object generally means the easier it is to see. The second factor is the *contrast* of that object with its background. If it stands out in color, say a black object against a white wall, it is easier to see than it would have been if it were a pale pink against the white wall. Another factor is the *time* of the seeing sensation. It takes time for an image to form in your brain. If you can't see the object long enough, perhaps you won't see it at all. Finally, there is *brightness*, which has to do with the amount of direct or reflected light which comes into the eye itself to create the stimulation of the rods and cones. We consider home lighting because it directly affects that last factor, brightness, which governs our ability to see things around us.

## REFLECTANCE TABLE

The government has published a reflectance table which shows the percent of light reflected from various colored surfaces. We include it here as Table 7-1.

If we want to have a color illuminated like light maple so it can easily be seen, then we need more light on it than we would, say, on a color like light blue. Thus we consider the light for a den paneled with light maple wood against a bedroom which might be painted in light blue, both being about the same size. The den requires more lighting.

## INCANDESCENT AND FLUORESCENT LIGHTING

We have examined the effects of color on ourselves and learned something of its production. You're probably wondering what color lighting to use in various locations. Generally we find

Table 7-2. Selection Guide for Incandescent Bulbs.

| Activity | Minimum recommended wattage[1] |
|---|---|
| Reading, writing, sewing: | |
| Occasional periods -------------------------- | 150. |
| Prolonged periods-------------------------- | 200 or 300. |
| Grooming: | |
| Bathroom mirror: | |
| 1 fixture each side of mirror--------- | 1-75 or 2-40's. |
| 1 cup-type fixture over mirror ------- | 100. |
| 1 fixture over mirror-------------------- | 150. |
| Bathroom ceiling fixture -------------------------- | 150. |
| Vanity table lamps, in pairs (person seated) -------------------------------------------- | 100 each. |
| Dresser lamps, in pairs (person standing-------------------------- | 150 each. |
| Kitchen work: | |
| Ceiling fixture (2 or more in a large area) ----------------------------- | 150 or 200. |
| Fixture over sink------------------------ | 150. |
| Fixture for eating area (separate from workspace)------------------------------- | 150. |
| Shopwork: | |
| Fixture for workbench (2 or more for long bench)-------------------------- | 150. |

[1]White bulbs preferred.

**Table 7-3. Selection Guide for Fluorescent Tubes.**

| Socket and wattage | Description | Where to use |
|---|---|---|
| **Medium:** | | |
| 30/70/100 | Inside frost or white. | Dressing table or dresser lamps, decorative lamps, small pin up lamps. |
| 50/100/150 | Inside frost or white. | End table or small floor and swing-arm-arm lamps. |
| 50/100/150 | White or indirect bulb with "built-in" diffusing bowl R-40). | End table lamps and floor lamps with large, wide harps. |
| 50/200/250 | White or frosted bulb. | End table or small floor and swing-arm arm lamps, study lamps with diffusing bowls. |
| **Mogul (large):** | | |
| 50/100/150 | Inside frost | Small floor and swing-arm lamps and torcheres. |
| 100/200/300 | White or frosted bulb. | Table and floor lamps, torcheres. |

incandescent or fluorescent lighting in homes. Sometimes on the outside we find colored lights, but we don't always find them inside. Does this mean that white light from fluorescent lights and the yellowish light from normal incandescent bulbs is what we should use? The answer is yes.

Bulbs may be obtained easily. They range from 15 to 300 watts. Some are tinted (soft light) types. Bulbs with the inside white finish, which is a kind of milky white coating, are preferred for most home uses. They produce a diffused soft light which is pleasant and permits easy sight. Some of the clear glass bulbs are used in fixtures and chandeliers for effect, but they can be hard on the vision if you have to look directly at them.

Tables 7-2 and 7-3 list recommended wattages of various bulbs for different activities around the home. Recall, however, that the light requirements may vary with the age of the person using the light. It has been found that older people may require as much as seven times the amount of light as a younger person.

The placement and position of the light source and the distance you are from that light source will also govern the amount of light necessary at the source for proper illumination of any object. It may be that the incandescent type of bulb will not be best for some of your activities. We visited a hospital recently and were amazed at the amount and types of fluorescent lighting which were used. This lighting was cool and restful on the eyes.

Most households use fluorescent lighting in one form or other. This type of lighting may require special fixtures, although you can find circular type fluorescent lights which are designed to be placed into table lamps under the shades. In determining the lighting requirements of your home, office, shop, condo or apartment, you will want to study Table 7-4. The placement of the light with respect to your work or viewing is important in the selection of the size (wattage) of the light used. Table 7-4 gives the generally accepted sizes and their location with respect to the person using that light source for vision.

## PROPER LIGHT PLACEMENT

One expert of the Duro-lite Home Lighting Institute has listed some considerations for the placement of light sources for various

Table 7-4. Selection Guide for Fluorescent Tubes.

| Use | Wattage and color[1] |
|---|---|
| Reading, writing, sewing: | |
|     Occasional | 1 40w or 2 20w, WWX or CWX. |
|     Prolonged | 2 40w or 2 30w, WWX or CWX. |
| Wall lighting (valances, brackets, cornices): | |
|     Small living area (8-foot minimum) | 2 40w, WWX or CWX. |
|     Large living area (16-foot minimum) | 4 40w, WWX or CWX. |
| Grooming: | |
|     Bathroom mirror: | |
|         One fixture each side of mirror | 2 20w or 2 30w, WWX. |
|         One fixture over mirror | 1 40w, WWX or CWX. |
|     Bathroom ceiling fixture | 1 40w, WWX. |
|     Luminous ceiling | For 2-foot squares, 4 20w, WWX or CWX |
| | 3-foot squares, 4 30w, WWX or CWX |
| | 4-foot squares, 4 40w, WWX or CWX |
| | 6-foot squares, 6 to 8 40w, WWX or CWX |
| Kitchen work: | |
|     Ceiling fixture | 2 40w or 2 30w, WWX. |
|     Over sink | 2 40w or 2 30w, WWX or CWX. |
|     Counter top lighting | 20w or 40w to fill length, WWX. |
| Dining area (separate from kitchen) | 15 or 20 watts for each 30 inches of longest dimension of room area, WWX. |
| Home workshop | 2 40w, CW, CWX, or WWX. |

[1]=warm white deluxe; CWX=cool white deluxe; CW=cool white.

tasks. He says that lights which are properly placed for such tasks as reading, sewing and putting on makeup can help to avoid eye strain and fatigue. You can also save on bills perhaps if you have been using a larger than necessary light placed too far away from where you are performing your tasks. Some of the recommendations follow:

■ When reading or sewing, locate your light so the shade is no less than 47 inches from the floor, for a floor lamp. The shade should be 38 to 42 inches above the floor for a table lamp.

■ The lamp's base should be in line with the shoulder and about 20 inches from the material being read or sewed.

■ If the lamp is attached to a wall, it should be behind and either to the right or left of the person doing the task.

■ For bed reading your light should be about 30 inches above the height of the mattress. If you use a lamp, place it about 22 inches from the reading surface. The bottom of the lamp should be about 20 inches above the lamp shaft and about 16 inches back from the center of your reading material.

■ Lamps on a dresser or dressing table should be selected in proportion to the furniture. They may have shades as small as 9 inches in diameter. The lamps' shades should be translucent enough to let the light pass through them. Choose lamps so that the center of the shade is about 15 inches above the dressing table top, at approximately eye level. You need some relatively high wattage bulbs.

## LAMPS AND LAMPSHADES

Table lamps may have different bulb requirements than floor lamps. When selecting a table lamp, have the shade at least 16 inches wide at the bottom, 9 inches wide at the top and at least 10 inches deep if the lamp is to be used for study, reading or sewing. It is best to have a bulb size of about 150 watts. If the lamp has a multiple socket, use three 60 watt bulbs. The bulb should always be located in the lower one-third of the shade when it is in place. You will find that many lamps will have the bulbs high—toward the top. This is not the best bulb location for proper illumination.

### Lumens and Footcandles

When you use a lamp for desk work or study, you need at least 70 footcandles of light over the work area. The footcandle is that amount of light falling on a 1 square foot curved surface when illuminated by one candle that is 1 foot away from the surface. We find our light bulbs' illumination qualities stated in *lumens* on the carton or box. To find the footcandle power when we know the Lumens, we divide the area to be illuminated (in square feet) into the lumen value of the lamp. For example, consider a 100 watt bulb which has a capability of 1600 lumens. It will illuminate a desk of size 2 × 3 feet or 6 square feet. Then,

$$\text{footcandles} = \frac{1600}{6} = 266.66 \text{ footcandles}$$

We have more than enough light for the task. What bulb size, in lumens, do we need for 70 footcandles? We consider a degradation factor of 10, so we add 10 percent or 7 footcandles to the value of 70. Degradation can be due to dirt or whatever which reduces the light.

$$\text{lumens} = \text{area times footcandles}$$

$$6 \text{ square feet} \times 77 = 462 \text{ lumens}$$

You will need a 60 watt bulb. In Table 7-5 we can see the relative lumen values versus watt values for light bulbs.

## Lampshade Colors and Light-Passing Qualities

Some lampshades currently being used are shown in Fig. 7-4. Each type has its application and location in a home, apartment or condo. The shades which pass light easily are generally chosen for the portable type of lamp that might be used for many different tasks. Desk lamps may have a more opaque (passing less light) type of shade, or they may have a kind of hood or shade which passes no light at all. This is to avoid uncomfortable brightness in your eyes which may come through the shade.

If the shade is to reflect as much light as possible, regardless of what color or texture the outer part of the shade may be, the

Table 7-5. Lumens Versus Watts for Incandescent Bulbs.

| Lumens | Watts |
|--------|-------|
| 10 | 80 |
| 25 | 270 |
| 60 | 840 |
| 100 | 1640 |
| 150 | 2700 |
| 250 | 4400 |

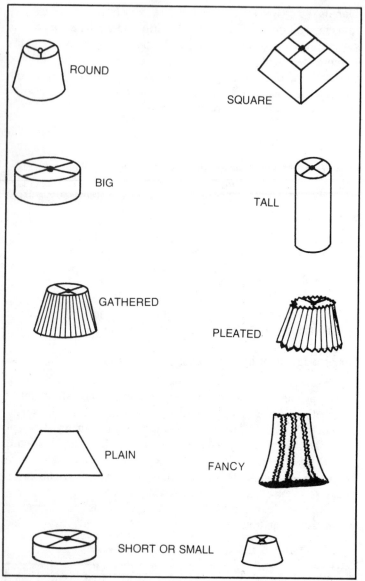

ROUND

SQUARE

BIG

TALL

GATHERED

PLEATED

PLAIN

FANCY

SHORT OR SMALL

Fig. 7-4. Various kinds of lampshades.

inside should be as nearly white as possible. In order to match room decor, the outer side of the shade may be of a neutral color or pale tint: off-white beige or a light gray. Darker colors are sometimes used such as burlap or chocolate browns. When thinking of room decor, try not to have too much contrast between the wall and lampshade colors. Vary the bulb size if you do not have three-way switches on the lamp, or dimmers for the wall socket, to get the best lighting effect.

### Diffusers

Fig. 7-4 illustrates the different shapes of shades. These shapes also affect the lighting produced. Straight or nearly straight lines are preferred by a majority of home owners instead of the extreme curves. Most people like the shades which give a wide dispersion of light downward and a small amount of light upward. Sometimes *diffusers* are an ideal way to disperse the light in a given room or area. Diffusers do not create glare and, to a large extent, they reduce shadows. The diffuser may be any kind of material through which the light must pass after leaving the bulb itself. You have seen many lamps, floor lamps especially, which have bowls surrounding the lamp bulb. The bowl is a diffuser and it is ususally of white, milky glass, although sometimes plastic is used. Some-times these diffusers are of colored glass or plastic which causes the emission of colors in the room. Then you can use that coloring to enhance your room decor in some instances. Some fixtures offer an "undershade diffuser," which is merely a plate inside the fixture just below the bulbs through which the light must pass. Some types of diffusers are shown in Fig. 7-5.

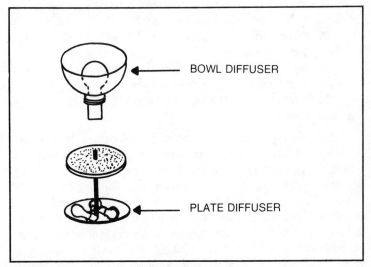

Fig. 7-5. Shown are bowl and plate eiffusers.

## CHANGING FIXTURES

Changing fixtures is one thing you might want to do around your home to change the decor, the general lighting effect and atmosphere. It is easy to change a fixture. Simply remove the screws, bolts or nuts holding the present fixture in place. Turn off the power to that socket before you make the change. Let it hang down on its electrical wires. Remove the connections to the cable wires, one at a time, being careful to keep the exposed ends far apart. Take down the present fixture and install the new one using the bars, screw holders or whatever mechanical parts are furnished with it. Connect the wires, making sure the caps cover all exposed metallic surfaces. Try it by turning the power on.

Finish the installation by pushing the wires back into the box and screwing the base of the fixture up tightly in place to the mounting bar, box or holder. Be sure you have turned the power off when you make the final installation. Sometimes you buy a fixture which does not have the mounting hardware furnished with it. Then you have to visit your local hardware store or the electrical department of a larger store, and get a kit of parts with which to mount fixtures. These kits are not expensive.

Here are some parts to keep in mind when you use new fixtures:

■ Incandescent bulbs should be no closer than ¼ inch to an enclosing globe or diffusing shield.

■ Top or side ventilation is desirable to keep the temperatures inside low and to extend bulb life.

■ Inside surfaces of shades should be of polished material or finished in white enamel.

■ The shape and dimension of a fixture should help support light efficiently and uniformly over the area to be lighted.

■ Plain or textured glass or plastic is preferred for enclosures or shades.

Examine Fig. 7-6 for examples of lamp placement for reading, studying and sewing. For shop work the lamps will probably be double fluorescent types mounted directly over the workbench at a height such that you cannot hit the light when working on any type object. The size of the bench light will be governed by the distance from the bench and the *age* of the person using the bench. Older persons need more light. Sometimes the fluorescent light is placed against the bench's back wall at a height of 3 to 4 feet so that it provides plenty of light on the work area but no glare in a person's

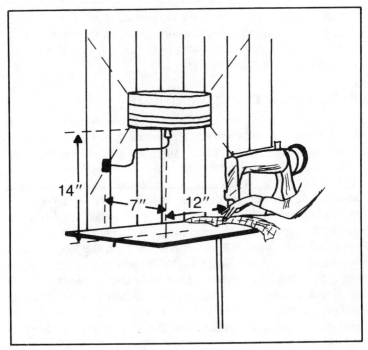

Fig. 7-6. Place the lamp properly for sewing.

eyes. We refer the overhead mounting for this type of light. Then you can stoop, bend and twist when working on something and not get the glare as you might if the light were mounted on a back wall of the bench area.

## SPECIAL LIGHTING EFFECTS

One of the big areas of improvement in lighting and the effect of lighting on decor is the use of *concealed lighting*. This may vary from subdued lights hidden in wall corners, pointed upward to give a fine, soft illumination to a room, to the complete side wall illumination of *valances* and *cornices*. A valance is a light placed inside, but not touching, a curtain or drape over a window. Light from a valance is directed both upward toward the ceiling and downward over the drape. The cornice lighting is over a wall and has no upward type of illumination. The lighting is concealed behind a front board and is directed downward over the wall. Its light is full, bright or soft and warm, according to the amount of light used. Fluorescent tubes are used for these long runs of light, and two examples of how they are installed are shown in Fig. 7-7.

A smaller cornice light system can be used in the shop or kitchen. By smaller, we mean a shorter length fluorescent tube. In a living room where a whole wall is lighted, it may take several 6 or 8 foot fluorescent tubes to do the job.

## INSTALLATION OF FLUORESCENT FIXTURES

Fluorescent fixtures have a "built-in" transformer and a ballast of some type in the usual case. Generally, all that is necessary is to mount the fixture in place using long bolts or screws. Then connect the two wires (black and white) to the appropriate wires from the switch or "hot" line cable. Some fluorescent fixtures have switches on them of either the pull-chain or toggle type. You won't need a separate wall switch to turn them off or on. You *do not* use standard type dimmers with a fluorescent type of light, so don't try to do it. They need full power directly from the line which is stepped up to the high voltage needed to ignite the inert gas used in the tube. When the fixture is in place and connected and the switch is off, put the tubes in by sliding each end into its socket and turning the tube roughly 45 degrees or a quarter turn. That should lock its end contacts in place against those in each end socket. Then it is ready to use.

Keep the tubes clean just as you should keep all light bulbs clean. Your bills rise and you get less light if you do not do this. If you want to find out about a dimmer which can be used with a fluorescent light, then contact your nearest General Electric Company office. They make them. Dimmers for incandescent lights are easily found in hardware stores, etc. These dimmers, however, may cause some interference with your radio or television expecially if they are on the same branch circuit. I suggest you check one out by connecting it. Before you put it in place in the box and seal it up, test it for interference to your hi-fi, television set and radio. It might turn out that you won't want to install it because of the interference.

## HIGH WATTAGE BULBS

Some lamp fixtures are designed for bulbs of 150 to 250 watts. Often these have porcelain type sockets. You should give serious consideration to the porcelain sockets which are "temperature safe," particularly if you have a small shade, a metal shade or any kind of shade which gives small ventilation around the bulb. Many home lamps use big shades and have the sockets mounted so that

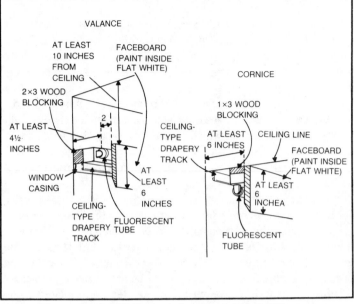

Fig. 7-7. The mechanics of valance and cornice light installations.

they get plenty of air circulation around them. Use the three-way bulbs of 50-100-150 or 50-150-250 sizes effectively for safe use.

Watch out for high temperatures around sockets which are of plastic, bakelite or similar material. Heat can cause the fusing, stripping and melting of wire insulation, the melting of the socket itself, and short circuits in the lamp and wire which can make the metal socket dangerous to touch. If you think the bases and lamp socket are getting too hot, a smaller bulb should be used.

## LOSS OF LIGHT ILLUMINATION

Did you know that the illumination levels of ordinary home lighting decrease with time of use or installation? This is sad but true. Lights deteriorate and can have their light producing capability reduced as much as 20 percent just due to age. The lights in the lamps will accumulate dust and dirt, etc., and can reduce the light producing ability from 10 to 30 percent depending on how long ago they were cleaned. It is suggested that a monthly cleaning of fixtures and bulbs is needed to get the most light consistently. When the lights get old, replace them if they don't burn out. The normal life span shown on a light bulb carton indicates about how long the bulb can produce light at the stated lumens value.

## MORE ABOUT LUMENS

Since the most used ratings of bulbs are lumens and watts, let us go over these characteristics again to be sure we understand their relationship. The lumen is defined as the amount of light which goes through 1 square foot of space, perpendicular to the light source, when that area is located 1 foot from a source that emits 1 candlepower of light in all directions. Examine Fig. 7-8.

We see how the surface of a round ball around the candle (light source) has one section which measures exactly 1 square foot in area and lets light pass through it. The amount of light from that source, through that ball which is exactly 1 foot away from the light source, is called a lumen of light.

We can expand this definition by noting that the area of such a sphere is $4\pi \, r^2$, where $\pi$ is the number 3.1416. From this little equation comes the relationship between candlepower and lumens. If the candlepower is to be computed, divide the stated lumen value by 12.57. Suppose you buy a lamp and use a bulb which has a rating of 1,500 lumens. If we make the division as previously indicated 1,500 divided by 12.57, we get 119.33 candlepower radiated inside the sphere which is 1 foot in diameter. The r value in the previous little equation is the *radius* of that sphere, and it has a value of 1 for our calculations.

### Light Intensity

Next we must concern ourselves with how the intensity of light decreases as we move away from the light source. The illumination of an object varies inversely as the square of the distance from the light source. We can state this as an equation:

$$\text{Illumination} = \frac{1}{r^2}$$

Note that is the distance from the object to the light source. If we have enough light (candlepower) to illuminate the object we are viewing at a distance of 1 foot away from the object, it will take four times as much light (in lumens) to give that same illumination to the object if we move the light 2 feet away. Notice carefully here that this means there is no other light from any other source, including no reflected light to help illuminate the object. In practice you will have reflected light to help illuminate an object, and the relationship of requiring four times the light may not hold true under all actual conditions. But you will have to have some increase in the light source intensity if you move the object of the illumination further away from a given light source.

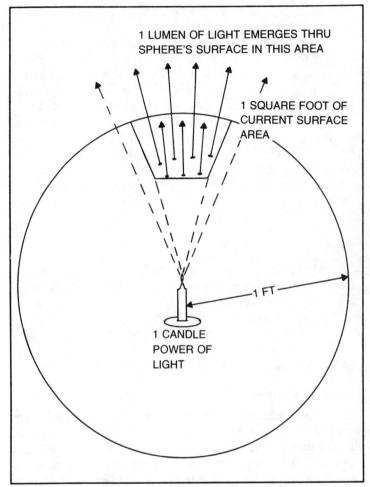

Fig. 7-8. This drawing will help you understand the definition of a lumen.

We have stated that 1 footcandle of light is that amount of light falling on a curved surface 1 foot away from a light intensity produced by one candle. This is also the same amount of light required for one lumen. So 1 footcandle of light requires one lumen of light per square foot. Thus, if we want to illuminate a bench area of 18 square feet to an intensity of 30 footcandles, and we know that 1 footcandle is 1 lumen, we will need 18 lumens to provide 1 footcandle. Since we want 30 footcandles, we then need 540 lumens (30 × 18) to give the required illumination to the bench. If we move the light source away, we will have to have more than that according to the inverse square rule. We leave that calculation to you as an intellectual exercise.

As a guideline, notice that 50 footcandles are generally required to read by. Sewing needs up to 175 footcandles. A kitchen area uses 30 to 50 footcandles. Seventy-five footcandles are needed in a bathroom for shaving with perhaps 30 for makeup. The older a person is, perhaps the more illumination is needed.

A 100 watt lamp bulb produces about 1,700 lumens, a 60 watt bulb generates about 850 lumens, and a 40 watt bulb produces about 500 Lumens. A 250 watt bulb, of the three-way type can produce up to 5,000 lumens. Examine the bulb carton next time you go to the store. Notice the relationship between the wattage size which is always stated and the lumen value which is frequently given.

## Meters

Perhaps you are wondering how to determine the illumination of a room in footcandles. There are meters calibrated in footcandles which you can buy or perhaps borrow to use. These might be found at photographic stores, for all photographers are very careful about light intensity and effects. Photographers may be good people to get acquainted with if you want to consider special lighting, without glare and "hot spots," in a room situation.

## FLUORESCENT LIGHT COSTS

Every fluorescent fixture has a sticker on it. It says that if you want to find the volt-amperes (watts) used by that fixture, you must multiply the sum of the bulb wattages by a figure of 1.25 or 2.5. The value of 1.25 is used for the rapid-start 40 watt fluorescent lamps, and the 2.5 figure is used with all circline and other types of fluorescent tubes. Although you may think that the fluorescent type light light is less costly than the incandescent type light, this may not be true. The kilowatt hours of electrical consumption will be based on the total consumption of the electrical current by the lamp. That includes the light-producing elements, the transformer efficiency and ballast consumption.

A fluorescent tube light which is rated at 60 watts actually will consume 150 watts of electricity, using the 2.5 factors. A rapid-start fluorescent light of the same wattage will consume 75 watts of electricity. This doesn't mean that you won't get good lighting which is cool and nice from the tube. Just be aware that you may not be saving as much on that electricity bill as you had previously

Fig. 7-9. This lovely chandelier would enhance the appearance of any home.

thought. Note that the light equipment may produce some heat which accounts for the increase in wattage.

## IMPROVING ROOM DECOR WITH LIGHTING

The fact that you can buy many different types of fixtures which can improve lighting and decor makes this facet of home wiring and lighting extremely interesting. Look at the fixture in Fig. 7-9, a chandelier which has small, candle type bulbs in it. It

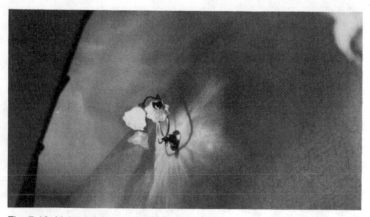

Fig. 7-10. Lighting improves a stairway's decor.

beautifully enhances the decor of this living room. Notice the long drop chain and cord.

You can find new ways to illuminate stairs and "dark areas" in the home by just changing the fixtures. In Fig. 7-10 a change from a small, unimportant light to the fixture shown added excitement and enhanced the decor of a stairway. In a home with stairs you may want to add some low level lighting at the stair level to insure that persons won't miss their step and fall. Subdued and concealed stairway lighting improves the decor of any home.

Sometimes it is possible to go through a wall from a receptacle outlet in an adjoining room in order to get the necessary electrical power for new light fixtures. Often you can go into the attic and send down a Romex cable between the studs to the wall location you have chosen for new lights. Sometimes you want a light on a wall and just cannot get the electrical power to it unless you either tear out the wall and patch it or use surface conduit which can be painted so it blends into the wall coloring.

Occasionally you can use spotlights of small sizes, 30 to 60 watts, in place of a fixture. These spotlights direct the light so that you get a better and more dramatic illumination of a hallway, stairwell or some dark area in the home. Spotlights can be used in swivel fixtures which are easily attached to a box in place of some existing fixture in the home. Various shades of colored spotlights can be obtained for more dramatic effect.

## OUTSIDE LIGHTING

You may want to have good outside lighting of a home, backyard or patio in order to gain full enjoyment from that area in an evening's activities. If the home you have was well planned, probably you will have outside outlets on the wall of the home. They should be weatherproofed with closing covers into which you can easily plug in outlet cords for the type of lighting and electrical devices you want for the area. Attachment to the electrical circuits in the home gives you 110 volts at the branch current amperage, 15 to 20 amps.

There are some types of outside lighting which work on a 12 or thereabouts magnitude of voltage. This type of system uses a step-down transformer which can be connected into the home branch circuit outlet. It supplies the lower voltage (at increased

Fig. 7-11. The conduit for this fixture is metallic.

Fig. 7-12. An outside light with a receptacle.

amperage for the same amount of light) to the fixtures and bulbs which are placed at various intervals around the area or structure to be lighted. Some of these low-voltage systems advertise that you can use their cords, which are waterproof and weatherproof, and just lay them on the ground or bury them underground without having any problems.

We don't recommend use of exposed cables anywhere underground or on the ground. We believe that the cables should be encased in plastic tubing well sealed at each end. The cable should be weatherproof and moistureproof as the code calls for. In most cases where the 110 volt underground runs are made, you should use rigid tubing of the metallic or PVC plastic type. Be sure that the ground doesn't shift or settle so that a bending strain is placed on the tubes. When you use this type of conduit and run wires or cable through and seal the ends, you have the best protection from electrical problems. Figure 7-11 shows a lamp fixture connected to a box used to pull the cable or wires through the conduit and make connections. The box is connected by screw type fittings to the rigid metallic conduit.

In a good fixture you will find a rubber grommet which goes around the bulb base. The bulb screws down tightly against it so no water can ever get into the socket. That is the proper type to use.

Fig. 7-13. A light with a metal hood for protection.

Fig. 7-14. A row of outside illuminating lights.

Sometimes you want to have an outlet in addition to the light so you can operate trimmers, barbecue lighters and televisions. In Fig. 7-12 you see how this can be done. The box is weatherproof with a rubber seal. This particular box uses a covered outlet section. There are two covers and each seals down with its rubber insert, tight, against the receptacle socket itself. If you need just one, you do not have to have the other exposed. If you need both outlets, then you can open the cover to the bottom one. Again, notice that a sealed box is used on an extender of rigid metallic conduit or pipe.

A good view of the metallic hood over the floodlight is shown in Fig. 7-13. There are no exposed wires at any place on this installation. The light itself is protected against accidental damage.

## Row of Floodlights

When illuminating a large area of the home, garden or patio, it may require many lights to get the amount and effect of illumination you desire. In this case you may use a row of lights, all connected by an underground piping or conduit system as shown in Fig. 7-14.

Because of the amperage demand, there are four 150 watt floodlights using 6 amperes of current. You may want to have a separate branch circuit for this row of electrical devices. Of course, as we already know, the code permits up to 10 receptacles (eight is best in modern practice), 15 amperes demand total for a minimum size wire in a branch circuit of 12/2, and a circuit breaker of 15 amperes trip value. That size circuit would easily handle this amount of lighting.

## Sloppy Outside Wiring

So much for the proper manner of putting in outside lighting. We won't say that you should use the 110 volt type lighting and outlet system or the lower voltage 12 to 24 volt system. Either will work nicely for you if you carefully follow the code for putting in such outside lighting and electrical outlet systems. Of course, if you plan to run appliances, you will need the 110 volt outlets.

Don't do a sloppy job of putting in outside wiring. See Fig. 7-15. Here a line was run through the wall from the garbage disposal in the kitchen to the separate garage. This line had to go underground in a plastic cable sheath. Instead of using rigid tubing to go through the wall, with fittings and boxes on each end, the person just ran an extension of the cable from the garbage disposal box through the "ell" and cable to the garage. As the ground settled outside, a pull on that cable sheath forced the "ell" out of the wall. You see the result in Fig. 7-15. Bugs can enter and the cable inside the sheath is exposed, creating a dangerous situation.

The other end of the cable system to the garage was just as bad Fig. 7-16. Running up out of the ground and into the garage without any fittings or fastenings to close the plastic conduit, the end is left open for rain and moisture to enter and settle on the electrical cable inside the sheath. The *least* that could have been done here would have been to run the plastic, flexible cable conduit inside the garage and fastened it there. As you can see from the hole size, this

Fig. 7-15. A dangerous outside connection.

was never done. We will never understand how this installation passed inspection.

So it may have been ground settling that caused the problem. We won't know the answer to that possibility. But we do know that if the job had been done properly, there would never have been any exposed cable at any time (Fig. 7-15). You want to make certain, if you put in outside wiring yourself or have it don, that it is done properly.

## Electric Eye Units

For the person who comes home after dark, there is no more comforting sight than to see the light on over the entrance or a light illuminating the grounds. This can be accomplished automatically with the "electric eye" control units available at most stores. These can be screwed into the light socket just as a bulb is put in place; then the bulb is screwed into its socket. The light is turned on and the "electric eye" part of the unit will sense whether it is day or night and close the circuit to the light.

## Timers

A second type of unit which can easily be adapted for this type of use is the small electric timer. The timer is motor-driven and easily available at most stores. You can set the on and off times for this unit. It will plug into a receptacle, not a light socket, unless you get a small conversion plug which converts the light socket into a receptacle type outlet.

If you just want the outside light for a couple of hours around twilight when you come home, and then off for the rest of the night to save money, this type of unit will do the job. You will have to get an extension cord to run to a light socket outside. Plug this extension cord into the receptacle socket provided on the top or side of the timer unit. Then you set two arms, one for on and one for off times. Be sure you differentiate correctly between a.m. and p.m.

The timer also can be used inside to turn lights off and on. It might be nice to have this set if you come home late to what is normally a dark house or apartment. Just plug it into a wall receptacle, set it properly and plug a lamp into it and turn the lamp on. Of course, the lamp won't light if the time doesn't correspond to the time for on that you have set on the timer. That is the way it should be. Normally these little timers have a rotating type of switch which you can turn halfway around to see if the light will work

Fig. 7-16. The open end of underground plastic conduit permits water and moisture to enter.

through the timer. Then you continue rotating this switch to turn the light off and make ready for the automatic timer control.

If you want to illuminate large areas at night automatically, you need a professional timer unit which you buy from your electrical supply store. If you have difficulty locating one, call your electric company and ask for help, and wire the timer on a separate branch circuit from the circuit breaker box and have the lights connected to it. You can set in about the same manner that you set the little portable timers. This one will be large and in a metal box mounted on a wall inside or outside the home. If it is outside, you can padlock the cover using the fittings which are a part of the cover and box. That, of course, should be done if it is outside.

For the single person who comes in after dark, it is nice to have an electric eye type of light over the doorway and in the garage area. Use one or several timers inside and set them to turn on various lamps at different times. If you have a portable radio, a timer can be used to turn it on.

## USING DIRECTED LIGHT

If you use a standard light bulb in a standard socket, you get room illumination. The amount of light falling on any given area may be only a fraction of the total light. Of course, we know that this light on a given area will come from direct light and reflected light from walls. But it may not be enough for reading, sewing or study. One solution is to use a larger bulb. Then you create more heat and, of course, there is an increase in electrical consumption which means a higher electric bill.

Perhaps you don't need a full room illumination. Consider using small spotlights, 30 to 60 watt types, which screw into regular sockets and focus the light into a relatively narrow beam over a distance of 6 to 8 feet quite nicely. These devices will provide plenty of light on a page or on sewing or study material while not illuminating the rest of the room as much. If you replace a 250 watt light with a 30 watt focused light there is a definite savings in cost. Let's see how much you will save.

## FIGURING ELECTRICITY COSTS

On your light meter there should be a kilowatt hour figure. It may be a number such as 6.8. This is the meter's calibration number. It relates the number of revolutions of the disc inside with the killowatt hours of energy, for example; 6.8 revolutions are equal to 1 killowatt of energy. Next, you will want to turn off everything in the house but one device, say the 250 watt light. Be sure it is the only device which makes the meter wheel turn. Next, count the number of revolutions of that disc per minute. There is a mark on the disc which you can use to identify a starting point and accomplishment of a revolution. Count the number of times that mark appears opposite a given place on the meter face for 1 minute. If you then multiply the number of revolutions times the kilowatt hour factor (6.8 in our example), you will get the rate of energy used in watts/hour.

You can call your electric company and find out its charge for electricity in kilowatt hours at your home. Since you have determined watts of usage, you now must determine kilowatts by dividing the watts by 1,000. Multiply the result by the cost per kilowatt hour figure you got from the electric company, and you have the cost per hour of operating that particular lamp. Repeat the process for the focused light (30 watts) and compute costs per hour. You will find the focused light much cheaper to operate. Do this procedure for any kind of electric device and find out what it is costing you to operate. If you cannot turn everything in the house off to make this kind of evaluation, then leave the device off and compute the energy rate of usage without it being on. Then turn it on and recompute the figure. You *should* get an increase in cost. That increase will be equal to the cost of operating that additional device only. If your meter has no kilowatt hour factor, call the electric company and ask what it is.

## READING THE ELECTRIC METER

Let's go through the reading of the dials on that electric meter. The formula for reading and determining cost from the meter is:

$$\frac{\text{Watts}}{1} \times \frac{\text{Cents} \times 0.001 \text{ Kilowatts}}{\text{kilowatt hour} \quad \text{watt}} = \text{Cents per hour}$$

As you can see, the various terms cancel to give us the value we want.

Suppose that your cost of electricity is 3 cents per kilowatt hour. Assume that you have a lamp which is consuming 300 watts of electricity all the time it is on. We insert these values into the equation as follows:

$$\frac{300}{1} \times \frac{3}{1} \times \frac{0.001}{1} = 0.9 \text{ cents per hour}$$

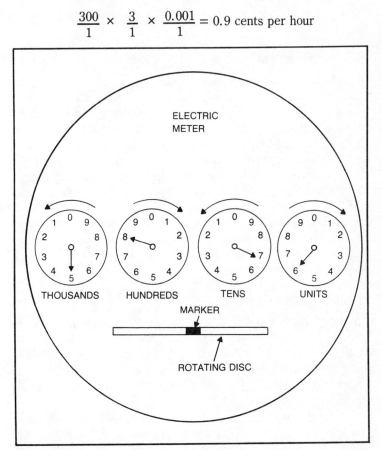

Fig. 7-17. Shown are the dials on an electric meter.

This works out to almost 1 cent an hour which is pretty economical. Refere to Fig. 7-17 and let's read the dials. Notice the direction of the numbers change according to whether you are reading in units, tens, hundreds or thousands.

We always start with the units dial, which reads 6 in this case. Then we go to the tens dial and read 7. The hundreds dial reads 8 and the thousands dial reads 5. When we put the numbers in proper order, we get 5,876.

If the reading hand happens to fall between two numbers you always select the lower or smaller number as the one to write down. Subtract this reading from the previous month's reading and you will have some idea of the electricity consumption you have had during the month. If you know the cost of electricity, as we have shown previously, you can calculate the cost of your electric bill.

## SAVING ON THE ELECTRIC BILL

The use of proper lights and turning off those which aren't needed is a good way to start saving costs. A second way is not to stand at the refrigerator with the door open wondering what might make a good meal. A third way is not to do the laundry until you have full loads. A fourth way is to turn back the water heater a notch or two if it is electrically operated. If it is gas operated, do the same and save on the gas bill. Turn everything off in the home and check the meter to see if the disc is turning. If it is, you might have an electrical "leak" somewhere. Use small floor lights of, say, 30 watt size to replace big bulbs (250 watts). You can think of some other quick ways to reduce the bill. The main thing is that you won't save a lot by any one action performed once or twice. If you do it every day, you will save more money.

## SECURITY LIGHT

Sometimes you want a light outside to illuminate a large area, but which can be controled from inside. The one shown in Fig. 7-18 is such a fixture. Notice how it can be removed easily by unplugging its line from the weatherproof receptacle socket. The two other lines on the bottom right of the fixture go to a control switch inside the building.

When your roof has an overhang, it is easy to go through it into the attic for the "hot line" as well as for the control line. Notice that rigid conduit is used to make a connection to the weatherproof

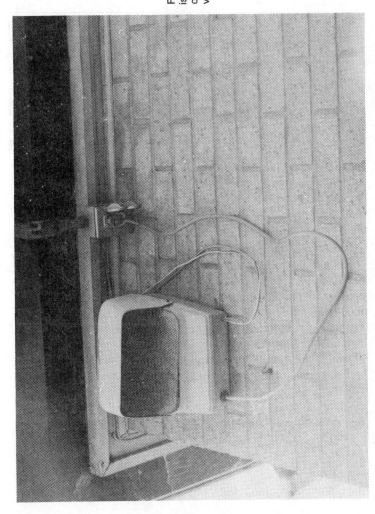

Fig. 7-18. This outside security light is plugged into a receptacle and has control wires separate from the 110 volt supply line.

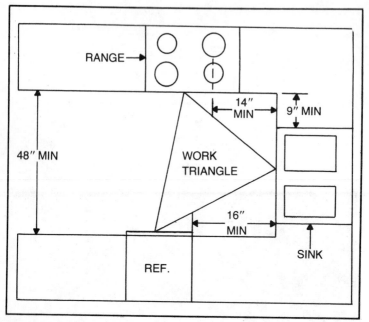

Fig. 7-19. The kitchen work triangle.

receptacle which is mounted as far back from the edge as possible to prevent moisture and water damage.

## KITCHEN WORK TRIANGLE

The heart of the kitchen is the *work triangle* as shown in Fig. 7-19. The sum of the sides of this triangle should not exceed 23 feet, according to a study by the Department of Housing and Urban Development. There should always be a minimum of at least 48 inches distance between appliances placed opposite each other on counters or self-standing. If there is a passageway between the appliances, then 30 inches between them may be sufficient. In an "L" or "U" shaped kitchen, the minimum edge distance between an appliance and an adjacent corner should not be less than 9 inches if the corner is a sink or 16 inches if the appliance is a refrigerator.

### Refrigerators

Refrigerators should be installed where the interior is accessible from the work triangle and the open door will not interfere with the traffic flow. In our present home, if you open the refrigerator door you cannot get to the outside door. That is one example of a very bad layout arrangement for appliances. But we cannot do anything about it unless we move the back door (impossible) or do without the icebox (also impossible). If you have a double-door refrigerator, the doors should swing away from the work triangle.

## Ranges

A minimum of 18 inches of counter space should be provided on each side of ranges, or 24 inches on at least one side of separate oven units. The range should be chosen with controls located at the front or side. It should not have any storage areas between the burners. This reduces risk of burns from    plattering grease, boiling water or steam, and also the risk of catching your clothes on fire when reaching over hot burners.

Range hoods are installed to remove heat, smoke, moisture and odors along with providing a light source immediately above the cooking area. The hood should have rolled edges and be installed so the cooking area is clearly visible for one standing in front. The optimum height for hood installations ranges from 56 to 60 inches above the floor. The depth of the hood determines which height should be used. If the hood is 17 inches or so deep, then the 56 inch height figure is appropriate ; if the hood is 18 or so inches deep, then the 60 inch figure applies. Wider hoods or deeper hoods may require different height levels. Exhaust from the hood should go outside and never into an attic or other unused space.

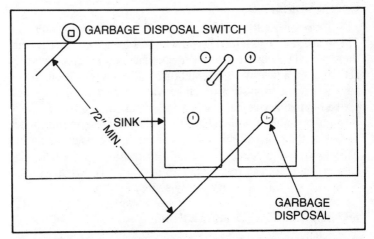

Fig. 7-20. The wall switch for a garbage disposal unit should be located a least 6 feet from the opening of the disposal.

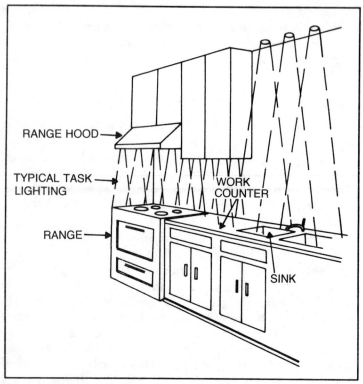

Fig. 7-21. Make sure the kitchen area has plenty of light.

### Garbage Disposals

There are two types of garbage disposals currently in use. These are the "batch feed" type which is loaded to capacity and then run when covered, and the continuous feed type which is activated by a nearby switch and runs until you turn it off. If you have a batch feed system which activates automatically, then install a momentary contact switch at a minimum of 6 feet from the disposal opening. You then have to move away to turn this switch on before the unit will run; then it continues to run automatically. This is a good safety feature. On the continuous feed, just be careful and don't get your hand into it while it is turning. Figure 7-20 shows the momentary safety switch location.

### KITCHEN RECEPTACLES AND LIGHTING

We have discussed the code requirements in a kitchen, but it may be wise to repeat some of those ideas here. An adequate

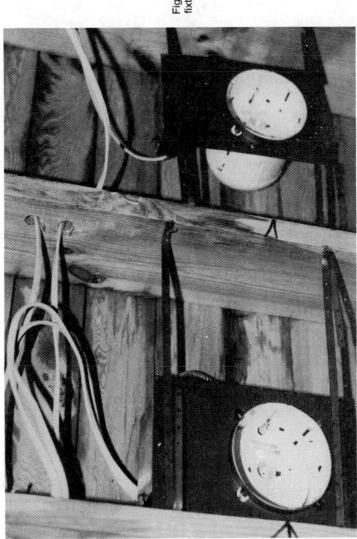

Fig. 7-22. Two recessed light fixtures over a kitchen area.

number of receptacles in the kitchen will make it safer by eliminating the use of "extension cords" and other makeshift methods of getting electricity to the numerous appliances found in the kitchen. The receptacles and switches should be located as far away as possible from any water outlet piping or water fixtures. When you do your kitchen planning, follow this general rule. Allow one receptacle for every 4 linear feet of work surface.

Notice Fig. 7-21 where the lighting is direct and plentiful. Of course, other lights will be placed over tables and for general ceiling illumination. Hanging and wall-mounted lighting fixtures should be placed no lower than 6 feet, 8 inches above the finished floor, except where installed above a permanent structure which could restrict traffic flow in the kitchen. You don't want any fixture or electrical device so positioned that a person can run into it and be hurt.

The kitchen fixtures for lighting should be planned to provide bright coverage for the entire room and the specific work areas involved. There should be a miniumum of shadows. This same concept will apply to a bathroom as well.

In Fig. 7-22 we see another example of the recessed type of lighting alluded to in Fig 7-2. This type of recessed lighting is often used over the bathroom wash basin and counter area. These lights may have plastic grills over them for ventilation purposes. They use porcelain sockets, and the wide space for ventilation around the bulb area permits large size bulbs to be used. The lights may also have an opaque glass plate which diffuses the light and makes the room much more comfortable. Notice how hangers are used to get the lights into just the proper position for maximum lighting with minimum shadow.

Of course, we don't want to imply that these types of fixtures are used only for kitchen or bathroom areas. They are also used in dens, family rooms and even in hallways and entrances. If you add a dimmer control to one or several of them, be sure the dimmer has sufficient wattage capacity to handle all the lights or bulbs you want to control.

# Rewiring Older Homes And Electrical Renovation

How old is old? When it comes to electrical usage and installation of electrical devices, outlets and lights in a home, old can be one or two years. After that length of time, unless the home was well planned in the beginning, you will find you are using more connections to the electric receptacles to run more things than you really have outlets to handle. You will probably be using the three-way receptacle adapters, or extension cords with three outlet connections, to electrify things in your home.

How do you expand what you have, if you have the necessity to expand your electrical wiring? That depends upon how old the house actually is. If it is a very old house, you might find wiring like that shown in Fig. 8-1.

This wiring is run along a corner of a ceiling on porcelain "knobs" or clamps which hold the wires away from the structure itself. Sometimes these porcelain knobs are useful when rewiring an older home if you do not have access to a space between floors or in a ceiling or attic. When modernizing, you should use surface wiring conduit or water-resistant cable (Romex) which can be stapled to the structure if possible.

## CHECKING THE SERVICE INPUT

Where do you start with such a rewiring job? The best place is at the service input. We know from our previous study that a service input should have a capability of 100 to 200 amperes. Two hundred amperes is about adequate for today's modern living re-

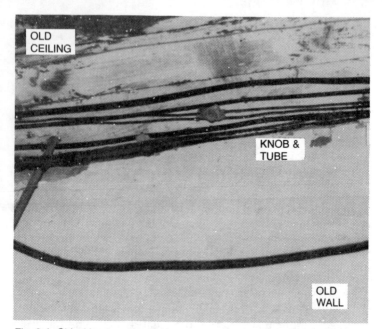

Fig. 8-1. Old wiring is supported by porcelain insulators and clamps.

quirements. We begin by evaluating the system to see what we have as an input.

You will usually find that the service input to an old house is far too small in capacity to permit having the lighting, outlets and electrical devices in the home which you might want for convenience. You may find that the electrical wiring is on the porcelain insulators and probably is about size 14 wire. Figure 8-2 shows one service connection to an older home which was being restored. Notice that the three wires from the electric company's pole are fastened to a simple strain insulator. Then they vanish into the home through the attic space. Inside they are fastened to separated lines of size 14 wire which feed all through the home.

It is this type of service connection which must be replaced in order to make way for some type of modern wiring in the home. You need the modern meter and circuit breaker panel arrangement to control the electrical input and make it safe and sound. If you want the home to appear in its original shape, color and style, you might have the electrical meter outside and concealed in some type of frame or box which is accessible for the electric company. Have the circuit breaker panel just inside where it is not visible from the outside and is not conspicuous to people inside. You do need a new service, and the electric company's representatives will be glad to assist you in determining what you need and how to best install it.

## WIRING AN OLDER LAMP FIXTURE

In our research for this book we examined many restored older homes with some modern wiring. Notice in Fig. 8-3 the long stemmed, beautiful fixture which once was a gas light fixture. It was kept and rewired for electric bulbs of a size and type which best matched the original brightness and style of the gas flame. In this home the high ceiling was restored. The full length windows of a size seldom seen today added charm and grace to the rooms.

It is almost always possible to electrically wire such fixtures so that the original decor of the homes is maintained. Simply replace the gas outlets with an electric socket. Wire all sockets in parallel and connect them to the stem wires which run into the attic and are attached to the main line Romex in a plastic or metal box. If a remote wall switch is desired, then you must find the space where you can run a cable down the wall between studs. In some cases you must use a surface conduit or cable near a door facing and get a cable down from the box to the switch location. The connection of the switch and light sockets is shown in Fig. 8-4.

Although we have shown the run from the switch to be a red or black wire which attaches to all the connections from one side of the socket fixtures, this line could be part of a white-black Romex cable which goes down to the switch from the light fixture box. The

Fig. 8-2. An old service connection to an older home.

Fig. 8-3. An electrically wired gas light fixture.

wire from the parallel connections to the light sockets then would run up the support pipe from the sockets just as the "hot" line does. It would connect to the black wire coming from the switch which would then be located inside the box. In this manner you just have a cable going down to the switch from the box. All connections to the switch and hot lines would be inside the box, and all lamp connections would be inside the fixture structure itself. If it is necessary

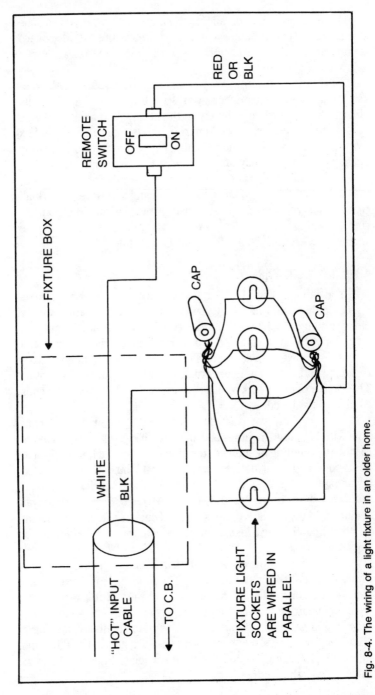

REMOTE
SWITCH

OFF
ON

RED
OR
BLK

FIXTURE BOX

CAP

CAP

WHITE
BLK

"HOT" INPUT
CABLE

TO C.B.

FIXTURE LIGHT
SOCKETS
ARE WIRED IN
PARALLEL.

Fig. 8-4. The wiring of a light fixture in an older home.

to run the fixture wires parallel to the metal supports, you can use small, painted metal clamps to hold the wires on the outside of each support arm so that those wires are scarcely visible from a position below the light itself.

This type of wiring is very common in restorations and renovations of older homes where you want to preserve and use the older fixtures which are already in the house. You must be aware though, that a new cable was run from the new circuit breaker box to the fixture box. You do not, usually, connect to the older existing house wiring which may be two lines run on porcelain insulators. However, if the load is not too large and the insulation is still good, you might be able to connect to such wiring and have it work all right. When you can't get to the wiring "runs" in an older home to replace wiring which is already there, you might be able to use the wiring if everything looks fine. But be careful.

## A WIRING NIGHTMARE

You will probably shudder at what you see in Fig. 8-5. There is a story behind this mess. A young couple found a rather lovely older building which they wanted to convert into an antique shop. It was of relatively sound construction as far as the interior and outside walls were concerned. The structure would need plaster replacing and paint, but in general it was sound. So they bought it. They had a plan for lots of little rooms in which various antiques would be displayed.

The building was a true brick-walled structure, two floors high with high level ceilings and ancient wiring. The first plan was to look the building over and determine what type of lighting would be needed and what outlets and electrical connections would be advantageous. Because the walls were thick and of brick, you could not run cables down through them. Much surface wiring was required. Many runs for various types of outlets and devices meant lots of connections to a fuse or circuit breaker box, as you can see in Fig. 8-5.

The electric company was called in to help determine the size of service needed for the building. They did their part well, efficiently and courteously. The buyers had the proper service entrance conduit, boxes and metering receptacles installed. Then they began the connections to that service themselves, using a National Electrical Code book as a general guide but without the help or assistance of a good master electrician. Figure 8-5 shows the result of their efforts.

Fig. 8-5. This method of wiring should be avoided.

## Analysis of the Wiring Job

The system works; there is no doubt about that. Circuits are protected. The proper number of circuits or branches and the proper size cables are used. But the method of installation leaves something to be desired.

This building was wired by making a "run" to the circuit breaker box as directly as possible. Junctures (ceilings and walls) were used as much as possible to try and conceal the wiring. Also, the buyers used much "direct" wiring—just going from a given room or electrical connection back to the circuit breaker box as directly as possible.

You probably feel it should be possible to cable up all that wiring and make things neater after all the "runs" have been made since it seems to be electrically okay. We won't argue that point. If you look over the illustration, you can see above the small box to the left a porcelain insulator with wires on each side of its diameter. That is not right. At the bottom right you see capped connecting wires exposed—they should be in a junction box. Much improvement in wiring needs to be done to make professional, neat and safe installation.

## Using PVC Conduit

When you have lots of runs like this and cannot go into the walls as professionals do when wiring new homes, it seems that you will have a big bundle of cables to tie together as best you can. You may get a large size PVC conduit (or metallic conduit of large diameter) and run the cables through it into the circuit breaker box. That will make for a neater installation at the eye and ground levels and be safer, because the cables won't be exposed to possible damage.

In Fig. 8-6 you can see how a large diameter PVC cable was used to get around an archway. When painted, it will blend in nicely with the wall coloring and will not be conspicuous. It also protects the cables around the edge so they won't droop or sag. In the background you can make out the ceiling drop from a surface wiring tube of PVC to a box for a fan. That box had to be very strongly supported in order to support the fan structure and tolerate the vibrations, which always occur due to slight unbalances in the blades.

We see in Fig. 8-7 how the plastic conduit is run down from the ceiling level to the wall ledge and to a box. Fittings are used on the conduit so the connection to the box is strong. From the box, then, the cables emerge and run on the framing molding in both directions to terminate in hidden outlets in the desired locations. The manner of making the electrical "run" is good. The installer could have used a square type surface molding conduit to better advantage than this round type. It would have been less conspicuous when properly painted. Circular conduit tends to leave a shadow underneath on both sides always, and it is usually more visible than a rectangular type conduit.

Fig. 8-6. Use PVC plastic conduit to route some electrical cables over a doorway.

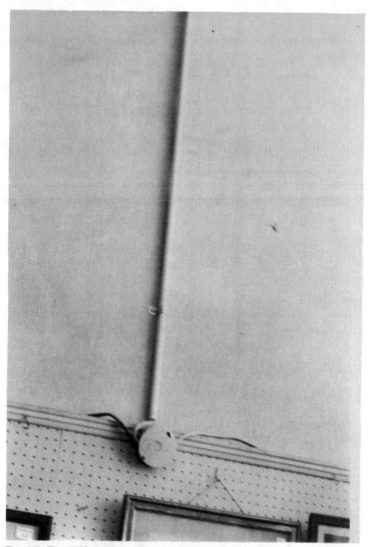

Fig. 8-7. The PVC conduit goes into a metal wall box.

## HIDING OLD WIRING

When you have old wiring or even new cables placed in the corner of the ceiling-wall juncture, you might undertake some carpentry to conceal them. The good electrician should also become a carpenter, especially if he is going to rewire old homes. In this case he needs to build a wooden tunnel which will hide and protect the wiring and can be later painted to match the room decor. The wood doesn't have to be very thick, possibly ¼ inch plywood or a similiar type of hardboard. You simply build a support side which is nailed to the wall and another which can be fastened to the roof. Then build a two-sided tunnel which will contain the wires and fasten to the wall support board and the ceiling support board or the ceiling fasteners. Figure 8-8 will give you some idea of this concept. Use your own imagination and igenuity to come up with other ideas of the same type.

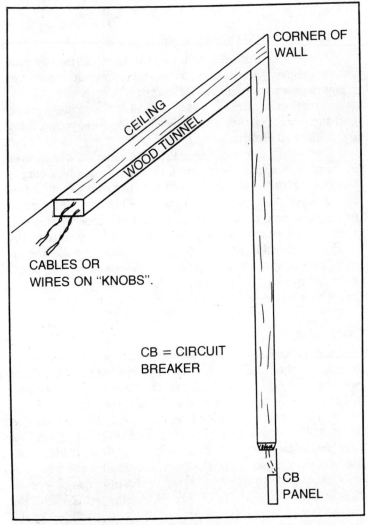

Fig. 8-8. Use of a wood tunnel for wiring.

There is nothing which says that a metallic tunnel couldn't be used, except that it might be bad if there was ever a frayed wire or cable or if moisture collected therein and affected the wiring. We like the wood idea better, though, as wood is an electrical insulator. Be aware however, that all wiring should be in good condition. The wires and cables should be separated and supported on insulators fastened to the ceiling or wall tightly so that all wire support is independent of the box, which merely will hide the scene.

## AN ELECTRICAL CARPENTER'S TASKS

An electrical carpenter is the man who does both the electrical and the carpentry work necessary when rewiring older homes or renovating some newer homes. When you have to remove a section of sheetrock or whatever to get to and install the wiring, and then have to fix things up after you get the electrical done, you are doing the work of an electrical carpenter.

An electrician friend of mine feels that more work is involved in wiring an old home. Sometimes, the friend notes, you can use hangers to get a box where you need it. If it's a ceiling light and the ceiling is of plaster, you can cut a small hole in it, run the hanger up into the space and work it until you get its ends supporting the box, which you fasten to a bolt. That bolt can move from end to end by resting on the inside of the plaster ceiling. When you make the box tight, you clamp the hanger against the plaster and it holds pretty good. My friend says if there are studs or joists you can reach, you might have to cut out more of the ceiling and fasten to them.

### Using Cable

This procedure is like we showed earlier in Chapter 6 for the mounting of the boxes between ceiling joists in new homes. There we saw how cables are used, but in rewiring an older home sometimes it is necessary to use BX armored cable which has wires in it. Greenfield conduit is another possibility. It is flexible and metallic, but you have to pull wires carefully through it. You may have to use Romex or other PVC conduit to get the job done. When using Romex, be sure to determine whether the "run" will need to have moistureproof or water-resistant cables. These cables should be used whenever there is the slightest chance of the cable being in a moist or wet area or subject to dampness of any kind. Figure 8-9 shows how the ceiling hanger might be positioned.

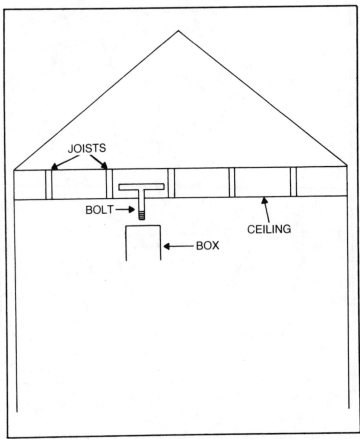

Fig. 8-9. The hanger is positioned through a hole in the ceiling.

Actually it is not fastened except by the box tension pressure when the box bolt is made tight.

### Going Through Flooring

The electrical carpenter may have to remove baseboards and run cables in the wall behind them. He may have to remove sections of flooring to get down underneath the floor and make his cable runs or fasten his fittings. If you have to take out a section of flooring, find the joists—sometimes you have to cut into the floor. Cut the boards at the joist juncture. Nail some support 2×4s to the joist so you can nail the floor boards back down again when you are finished, unless you make a trap door there. Then, too, you might have to add side supports as shown in Fig. 8-10.

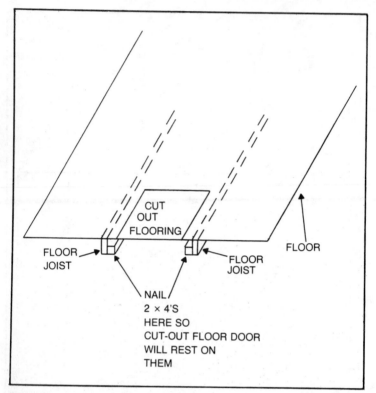

Fig. 8-10. Go through flooring carefully.

As for other drilling to run cables or wires, refer back to Chapter 6. While rewiring was taking place, we found the situation shown in Fig. 8-11. You can see how the holes were drilled so the Romex cables could be pulled through into the crawl space underneath the house and there routed to the next terminal or device point. The cables go upward to a switch and across to a receptacle box which is nailed to the heavy stud. You will note the similarity between this kind of cable "running" and that for the new homes which we have illustrated in Chapter 6. By carefully studying Chapter 6, you will be well prepared to work on renovation wiring.

## REPLACEMENT OF SWITCHES

Depending on the home's age you may find that you will be in for quite a task in replacement of old switches, even of the porcelain variety as shown in Fig. 8-12. While there may not be anything wrong with these switches, they are old and may fail. In planning

Fig. 8-11. Cables go through the floor, and a new receptacle socket is mounted in an older home.

317

Fig. 8-12. These two porcelain switches should be replaced.

the electrical remodeling you will want to replace the switches and all wall receptacles. The wall receptacles tend to have their plug holes enlarged by use. The contact metal inside gets worn away and sometimes bends so that very inferior and dangerous electrical contact maybe a consequence. You may find that the wires going to these receptacles and switches will be of the two-wire variety, coming through braided nonmetallic sheaths and porcelain tubes sometimes mounted on porcelain standoff insulators. You then wonder about whether to remove this old wiring.

It is best not to remove it. Just disconnect it and leave it in place. Removal takes time and may involve opening much more wall or ceiling space than you should. The replacement in effort and money to get the walls and ceiling back properly may be more expensive than it is worth to replace and remove the old wiring. After all, that old wiring won't hurt anything if it is not connected anywhere to any electrical source. Then you can pull in the new plastic Romex, run it inside the box easily when you push out the old wiring, and then insert your new switches and receptacles (Figs. 8-13A and 8-13B). Connect them to the new Romex, BX, or the wires coming through Greenfield flexible armor cable.

## NEW FUSE OR CIRCUIT BREAKER PANEL

Sometimes you will find that the old system for fusing and protecting the wiring in a home will be a big "knife" switch and a couple of fuses, one for each line. This was pretty standard some years back when electrical demands were simply for enough "juice" to power a few 40 watt light bulbs. In Fig. 8-14 you see how some new wiring required a new switch panel. Because the old wiring seemed sound and did not have too much of a load on it, it was kept in use. The new switch box just handled some additional outlets and lights and was a box type which could be closed. It still has two fuses inside and a "knife" switch. Fuses gave protection against excessive current demand.

When checking the old wiring, do not just see if the devices attached to it work or that electrical current can be obtained from the old wiring at all light outlets. Check to make sure the wiring is safe. See if the insulation is crumbly, frayed, broken or cracked, or

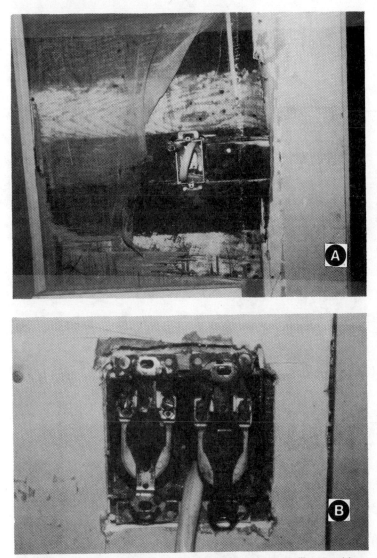

Fig. 8-13. Making connections with Romex. (A) Romex that will connect to a new switch is coiled in a new box. (B) Romex exiting from the old box will connect to new switches. Sometimes the old switches are used with the new cable.

Fig. 8-14. An old knife switch, fuses and a new circuit to a new, enclosed knife switch box.

Fig. 8-15. Old timber must be replaced as well as things of an electrical nature.

Fig. 8-16. This old light fixture has been rewired.

if any wires are touching each other. If so, replace them. Many a fire has resulted because this old wiring was placed where the two conductors crossed each other and touched. Then the insulation got old and broken and finally the wires themselves made contact. It's always best to replace old wiring.

## ELECTRICAL WORK FOR A HOME BEING RESTORED

It is popular today to obtain an older home which has some historical value and then restore that home to its original condition. The electrical part of the restoration may vary considerably. It may entail just replacement of the wiring so it is not dangerous. In some cases a slight expansion of the old system may be needed to properly display the antiques contained within the home. That requires additional wiring and careful placement of perhaps concealed lighting or proper color. There may be come delicate wiring required for light fixtures which may have been electric or gas-light types (Fig. 8-16). Wall switches near doorways, for an inconspicuous switch and concealed wiring from the light fixture in the ceiling down to this siwtch.

The type of components which are used in a restoration may be different than those used in a modernization of an old home. The purpose is the same, but the appearance and location may be an entirely different matter. We visited one old home which has been restored and were delighted when we visited the "children's bedroom and bath area. In the children's bedrooms we found all wall light switches no higher than 3 feet from the floor, so the children could reach them easily. In the bath, special light provisions were made with switches within easy reach for children. Careful consideration was given to safety by location and type of device.

One feature which would pose a challenge for any good master electrician, let alone a do-it-yourselfer, was the modern light panel switchboard in the restored and renovated kitchen area (Fig. 8-17). The switch could be used to turn off or on any light in the house, no matter where it was located. The switch panel had small indicator lamps to show which lights were on and which were off. This is a nice feature if you plan to have a modern home which has a light system that can be operated by a central computerized switching

Fig. 8-17. A modern light switching panel for all the house lighting. Indicator lights show which room lights are on or off.

system. If you leave home for any length of time, you could set the computer (or time-operated switching device) to turn on lights in rooms at various times while you are away—at night of course. The police love this idea because it deters burglars. But all the wiring must be planned for such control. We will discuss this in more detail in the chapter on home alarm systems.

## PLANNING A NEW ELECTRICAL SYSTEM

There are many aspects of redoing an older home which must be taken into account even as you begin the planning and layout of a new electrical system. As we have indicated, new wiring will be necessary. Probably new switches, receptacles, circuit breaker panels and a new service from the electric company will be needed.

When you look over the house to make the room layout, you should make a drawing much like the blueprint we examined earlier in Chapter 5. Show all the existing electrical installations. You can adjust and modify the plan so that when you start to work you know exactly what you are going to do. Also, plan on how the cables will be run. You can also, from your plan, discuss the cost of supplies and parts with an appropriate dealer. If the cost is too much, perhaps a replanning of the new electrical system design will be in order.

You can do a little work at a time as money and time are available. Continue the work until you finish the job. That, of course, could apply to the whole job of renovating, remodeling or modernizing the structure. That way you will get a finished product which is what you really want and not have a product which has had "corners cut" to save money and time. Call in a professional if you experience difficulty.

### Kitchen Needs

If you are planning to live in the home as soon as possible, perhaps it is best to start in the kitchen area with the electrical and other renovating work. The kitchen will be the most complex electrical layout of the home because of the appliances. The types of electrical installations—freezer, icebox, microwave, heavy duty outlets, will require many cables and boxes. Also, the circuit breaker box should be located somewhere close to this area if possible. You will want to plan for an oversize circuit breaker panel so you can have some room for expansion later.

### Improving the Traffic Pattern

Look through the structure and plan for new doorways, and hallways to improve the "traffic pattern" of people moving around in the home. This will affect the lighting of the other rooms. The use of three-way switches in halls, on stairs and in basements will become important. Carefully examine the structure to see if the wood members are rotted from normal decay, water or termites. The members must be examined if you plan to attach electrical connections and boxes to them. The floors must be strong.

## Water Systems

Water systems and grounding of electrical wires to these systems must be electrically and mechanically sound. You must be sure there isn't some metallic piping inside the home which then terminates into a plastic section at the house entrance which can effectively insulate that metal piping from connection to the ground. If this is the case, you must add ground rods to be sure the water system in the home is properly grounded. If the water system is old, leaky or rusted and you can't get a good electrical connection to the pipes, perhaps you should consider some new water pipes. Copper ones are best. Be sure all of them are connected electrically and grounded.

Sometimes one has a tendency to see a pipe and connect to it for a ground system. It may be a gas line pipe. That is not the best ground. The code says to use a water pipe in a water system.

## ATTACHING RECEPTACLES IN DIFFICULT PLACES

When you have a wall in which you want to place a receptacle or a switch, and there is no stud right there, you might have to cut into a plaster-lath material to make the fitting. Here you have a couple of choices. You might use toggle bolts through parts of the lath (metal or wood) and plaster to hold the box, or use a hanger mounted in the wall. You can also remove a large section of the wall to find the studs and put in a cross member of 2×4 for the box mounting screws, nails or whatever. Some say that the use of the newest "stud finders," which are sensitive magnets mounted so you can run them along a wall at the base and find the nails where studs rise up, are very effective at locating studs. You might also find studs by taking a thin nail and making little holes in the wall spaced about ½ or 1 inch apart. When it becomes hard to insert the nail beyond the plaster depth, you have probably found a stud.

## THE TEMPORARY SERVICE

When you rewire an older home, you will need some electrical outlets to connect drills, saws and other electrical devices for work

Fig. 8-18. Temporary service connection to an underground conduit provides power to run electrical tools.

in the house. You will have disconnected the older service input or had the electric company do it for you. Now you need to ask them to provide a temporary service so you can continue the renovation. If you are working on a newer home, they can probably use an underground extension as shown in Fig. 8-18. I like this arrangement because it is high enough to keep lines clear from the ground.

326

It has a covered receptacle panel to keep rain and moisture out and can be supported on a strong 4 × 4.

Often the electric company can provide a temporary service by connecting to a nearby electric light pole. If you will provide the 4 × 4 support for it, they will run the lines down to a similiar arrangment as that shown in Fig. 8-17. In this case the feed is overhead and not underground. If you want to dig a trench from your outlet location to the electric company's pole, put good PVC plastic conduit in the trench. Pull the service wires through it so the electric company only has to make a connection at their pole.

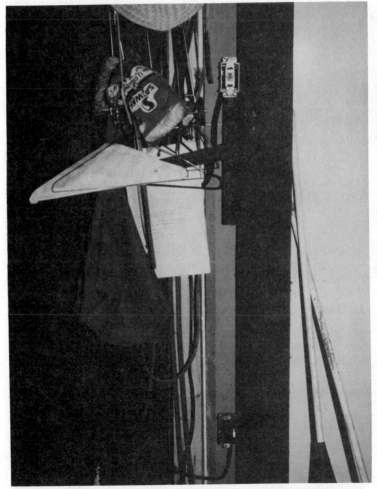

Fig. 8-19. Note the switch and receptacle near the housecleaning items.

## MORE ON ELECTRICAL CONNECTIONS

Figure 8-19 shows a switch and receptacle installation amid some housecleaning items. The stud here is an 8×8 structure which gives solid support to the ceiling. Nailing on a small piece of 2×4 provided a "perch" for the switch outlet. The receptacle outlet at the bottom was simply nailed to the big stud. Examine Fig. 8-19.

The installation of the sheetrock or gypsum board walls will not be done until the last item is installed and working satisfactorily. Then the walls are put in, sealed and painted or wallpapered. The job is finished by doing the floors in whatever finish and decor is desired.

Let us again show you the wires which come from the cables out of the box and are connected to the switch or receptacle. See Fig. 8-20.

Here you see the white, common wires, the black hot wire and the bare grounding wire. There is plenty of room on the box for the fixture. You can see multiple studs. Sometimes in renovation of older homes it is necessary to put in new studs. Nail or bolt them to the older ones in order to increase the strength of the wall. You do not necessarily remove the old studs unless dry rot or termites are evident.

Wiring a multiple element box again requires that you connect all white wires together and all ground wires together. Then the black hot wires are connected individually to each terminal of the switch or receptacle. A lead from the juncture of the white wire connections is also made to the second terminal of a receptacle and never to a switch. If a wire is connected to the second switch terminal, it should be a black, red or blue wire. Examine Fig. 8-21.

## REWIRING LIGHT FIXTURES AND LAMPS

If the fixture was a gas light, you have to use some ingenuity and imagination in routing the electric wires from the base (ceiling support) of the structure down to the lamp light position. There, fasten an electric light socket. In most cases you can feed some small, well insulated plastic-covered wires down through the hole which passed the gas. Be sure to use as large a size wire as possible, at least size 18. That is the size used on most extension cords.

The insulation should be thick and tough. Try to strip an end from the wire. If it is easy to strip and the insulation seems brittle and breaks easily or is very soft and stretches thin easily, beware of it. When you're satisfied that it's adequate, feed it down through the gas hole and out to where the socket will be attached.

## Socket Work

Sockets can be found which have a screw type fitting at one end. The fitting can be placed over some small pipes or be screwed into a threaded fitting at the end of the gas pipe. Sometimes you have to use a pipe *reducer*, which threads onto the gas pipe and has

Fig. 8-20. The receptacle or switch wires are exposed.

Fig. 8-21. Wiring switches in a box.

at its opposite end the right size fitting to screw into the socket. Sometimes you may have to just place the socket on the end of the gas pipe feed and tighten the screw of the socket fitting tightly against the gas pipe.

After affixing the base of the socket, you "break" the socket joint, made for the purpose, so you have access to the interior of the socket. You then connect the two wires carefully to the terminals, making sure you don't have any exposed wire showing or any loose strands making a connection with a metal part of the socket. We like to use a solder "dip" on the ends of the wires to make them solid and prevent strands from getting loose under the screw head and later causing trouble. Be careful not to use too much heat too long as the insulation will melt.

Slip the socket's insulated jacket over the device and put it back in place in its base. Pull your wires up through the pipe, carefully and easily, so there is no loose slack anywhere. Then you can connect the base end of the fixture to the ceiling box and the wires to the switch wires which were previously run into the box.

### Multi-Light Fixture

Sometimes you have to wire many lights in parallel if the fixture is a multi-light type. In this case you may have to take it completely apart to run the wires in a concealed manner. You need much care and patience in taking the fixture apart, putting in the wires and then getting everything back together again. The caps, shown again in Fig. 8-22, possibly can't be used because they take up too much room. Then you have to tape the wire ends. Be sure the tape is plastic electrical tape and not the cotton type tape which is sometimes used in electrical applications. Use several layers of the plastic tape. It sticks well and will last a long time. If you can use caps, like in Fig. 8-22, then do so.

Look at the lovely fixture in Fig. 8-23. Run the wires down the chain and connect them to the parallel "ring" of wires going to all the lights. Be sure to use a wire with an insulation color that blends into the fixture color.

## Three Switches In One Box

Often it is advisable to have up to three switches located in the same box, but controlling lights in different places. In Fig. 8-24 we show how these switches are wired.

The "hot line" comes in and has its black wire connected through a box to one fixture wire. The other fixture wire goes directly to one side of its control switch. The other terminal of each switch is connected to the common white wire from the "hot" line. Notice that switch terminals A, B and C are connected together in the box. Then one of these terminals is connected to a white common line. This same idea can be used for two or four switches which are located in one box.

Although we have shown for clarity a single wire going to the fixtures from the switches and one wire connecting to the white "hot" common cable, you may use Romex cable (which has has two wires in it) on runs from the light fixture boxes to the switch box. In

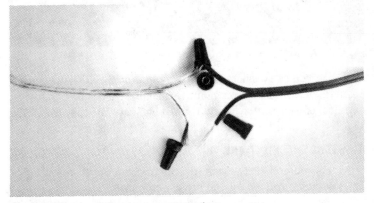

Fig. 8-22. Use insulating caps on connections.

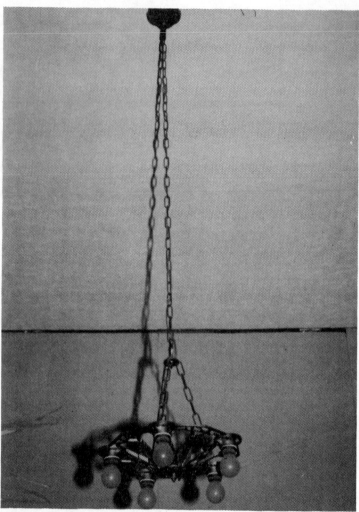

Fig. 8-23. An old light fixture with a chain support.

this case you can connect all white wires together. Then connect them to the white, hot common wire and make a connection to one terminal of each switch from the common white wire junction. The black wires will then go to one side of each light fixture.

## LIGHTING CHECKLIST AND TIPS

Ask yourself the following questions about the electrical lighting you have planned for the renovated home.

■ Can you light your way as you go from room to room?

■ Can you switch lights from each doorway in rooms with two or more doorways?

■ Can you turn on stairway lights before you go up or down the steps?

■ Can you light the front hallway and the living room as you come into the home?

■ Can you control the carport or garage light from the house? Do you have an indicator that the lights in the garage are on

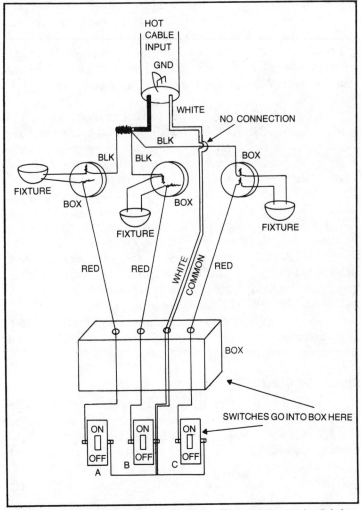

Fig. 8-24. A method for wiring grouped switches. Each switch controls a light in a different location.

or off if you can't see them from the house (pilot lamps over the switches)?

■ Can you control outside floodlights from inside the house?

■ Can you dim the lights in the dining room or bedrooms?

■ Can you dim the living areas for a change in mood and atmosphere?

These tips, of course, can apply to a new home as well as a renovated home.

There are some other tips which might be important when you redecorate or renovate a home. Finish the walls in light pastel colors and the ceilings in white, near white or a pale tint. Low gloss paint on walls and ceilings helps diffuse light and makes the lighting more comfortable. Use sheer curtains, or draperies, in light or pastel tints.

Add portable lamps for better balance of room lighting and plan plenty of outlets for them. Install structural lighting in living areas where there is only one ceiling light or no ceiling lights. Eight to 20 feet of wall lighting will add a feeling of space to the average size room and make the lighting more flexible.

Get bulbs with higher wattages if the ones you presently are using have low wattages. Be sure you do not exceed the wattage rating of the fixture containing the bulbs. A minimum of 150 watts on bulbs is very nice if you have three-way sockets. If you find that more than 150 watts of light are needed, then consider fluorescent lighting and change your fixtures accordingly. It is more efficient electrically to use one large 150 watt bulb than it is to use three 100 watt bulbs placed in different locations in a room.

Consider replacing outmoded fixtures, with modern, shielded ones. Sometimes you can come up with a nice blend of the old and the new in a renovated home.

## THREADING CABLES

You have seen how cables of Romex were threaded through holes, pulled over beams and stapled against studs and joists to keep them in place. You probably will need some help pulling the cable through all the places it has to go. You want to be careful not to exert too much force getting that cable through difficult places. You *can* pull the insulation right off it. Make your runs relatively short.

# Electricity
# Costs and Motors

9

We have to spend money for electricity. But imagine how it would be in our modern world if we had no electricity. Everything would grind to a stop. We tend to take our electricity for granted until an electrical storm or other event temporarily cuts off the electrical supply.

It's no secret that we live in a world of changing economics. As inflation and cost of living figures rise to haunt us, we need to consider costs of everything, including electricity, to see where we might save some money. Also, we need to evaluate the future carefully and conserve energy as much as possible so future generations will have plenty of electricity. While we help ourselves economically, we are also helping ourselves and others in the future. After looking at electricity costs, we will turn our attention to motors. Motor-driven appliances are some of the biggest users of electricity.

## ELECTRICITY AND HEAT

Unfortunately we have not yet discovered how to produce light without some heat. It's true that we have fluorescent lights which produce "cold light," but the transformers which make them work produce heat. In the summer this heat is wasted product; in the winter it may be helpful in reducing some direct expenditures for heating fuel. It has been estimated that in the summer much air conditioning cost is directly related to removal of the heat generated by lighting or operation of electrical devices. A home which has only fluorescent lighting will generate an estimated 50 percent less heat from this source than will be produced from incandescent bulbs which produce the same amount of light. Even if the relative cost of the two types of lights are nearly the same, the reduced cost of air conditioning when using fluorescent lights may be a large factor in paying less on that monthly electricity bill.

## THE HEAT PUMP

Units that produce both cooling and heating effects are called *heat pumps*. They pump the heat out of the house in the summer and inside the house in the winter. The units cost about twice what an equivalent air conditioning unit costs. But they do twice the job. Heat pumps look like an air conditioner. All you have to do to reverse the cycle is to adjust a thermostat or move a switch.

As of this writing, a heat pump for an average size home of 2,000 square feet costs $2,500 to $3,000. These units are about 50 percent more efficient than an electric furnace. Compared to an electric furnace, the heat pump will reduce electricity costs as much as 60 percent. Consult with your electric company and get a good estimate of costs and savings for your own particular home.

## ENERGY EFFICIENCY

We hear so much about the EER or *energy efficiency rating* of air conditioners. An air conditioning unit is rated in "tons" of air conditioning needed to properly cool your home. For example, one unit might be rated at 4 tones and another at 8 tones. We will see what this means later.

You must be careful when discussing the EER of an air conditioning unit. The EER can vary according to the temperature level being used as the basis for the rating. You must know what temperature is the rating calculated or measured. It could be that the EER is based on a temperature of 90 degrees fahrenheit. You need something around the 70 degrees fahrenheit for comfort.

Try not to buy a cooling system which provides more cooling than you actually need in your home. Most new homes are designed so that the air conditioning is just marginal—not much to spare—so the builder can save a few dollars. He may or may not pass on this savings to you. Without defining as yet the meanting of "ton" as related to air conditioning, we will give you some figures to use as a general basis for air conditioner size. Remember, this is not the EER. That rating applies to every unit after you determine how large a unit you need. The higher the EER—hopefully at around 70 degrees—will give you the least cost in operation.

A general rule of air conditioning is that you need 1 ton of air conditioning for every 400 square feet of living space. Cooling capacity is measured in British thermal units (BTUs). One BTU is that amount of heat needed to raise 1 pound of water's temperature one degree Fahrenheit. Five thousand BTUs is about the right size for one small bedroom. It is estimated that about 10,000 BTUs of cooling are needed for a den or family room. Now we can begin to relate costs of air conditioning to EERs. See Table 9-1.

## Tons

We give you now a definition of "tons" as the word relates to air conditioning: There are two definitions. The first says that 1 ton of air conditioning is 400 cubic feet of conditioned air per minute. This is somewhat general. The second definition of Ton is the cooling capability of 12,000 BTUs in a given time, perhaps an hour. The heat in a BTU is approximate to the amount of heat produced by burning a wooden kitchen match to ashes. In other words, 12,000 BTUs would be equal to 12,000 kitchen matches burned completely in a given air space. One ton of air conditioning is needed to remove the heat for each 3500 watts of lighting.

The EER is a number obtained by dividing the watts/hour into the number of BTUs of cooling obtained in that same time. Many air conditioner models have the EER stated on their label. Remember to ask at what temperature the ratings were obtained or over what range of temperatures. It costs more to cool down from, say, 60 to 55 degrees in many instances than it does to cool down from 100 to 95 degrees.

Table 9-1. Relation of Air Conditioning Costs to EER.

| Cost per Ton per Year of A/C | EER Rating |
|---|---|
| $108.00 | 5.5 |
| 99.00 | 6.0 |
| 91.00 | 6.5 |
| 85.00 | 7.0 |
| 79.00 | 7.5 |
| 74.00 | 8.0 |
| 70.00 | 8.5 |
| 66.00 | 9.0 |
| 63.00 | 9.5 |
| 59.00 | 10.00 |

**Table 9-2. Costs for Operating Various Electrical Items.**

| Item | Watts | Cost |
|------|-------|------|
| Attic Fan | 450 | 1.4 cents/hr |
| Portable Fan | 150 | 0.5 ¢.hr |
| Window Fan | 190 | 0.6 ¢.hr |
| Elect. Furnace | (variable) Ave=15 KW | $14.90 Kw/Yr. |
| Hair Dryer | 380 | 0.3 ¢ Hr. |
| Portable Heater | 1000 | 0.22 ¢ hr |
| Clothes Dryer | 5000 | $1.66 Kw/Hr |
| Iron | 1000 | 0.36 ¢ hr |
| Auto. Washer | 500 | 0.30¢ Hr |
| Sewing Machine | 75 | 0.3 ¢ Hr |
| Vacuum Cleaner | 630 | 0.12 ¢/hr |
| Water Heater | (variable) | $14.90 Kr/yr. |

## Operating Tips For Air Conditioners

Here are some operating tips regarding the use of air conditioners. In hot weather, use high fan or blower speeds. If the weather is humid, use a low fan speed to get out more moisture. Don't lower the thermostat when it becomes uncomfortable due to humidity; normally that doesn't help much. Keep shades closed so sunlight won't heat the house. Don't open the door more than you have to and keep windows sealed. The air conditioner may be the largest electrical cost item in your home, unless you use electrical heating. Conserve it by providing a climate in the home where it works as little as possible.

## COST OF OPERATING VARIOUS ELECTRICAL ITEMS

The cost of operating electrical items in your home depends upon the time, place and amount of usage. Table 9-2 will show some relative costs. They will serve as a guide; you can get the figures for your area by calling your local electric company.

You will notice that the electric furnace, clothes dryer and electric water heater have the highest costs. Reducing the temperature on the water heater by 10 degrees can make a significant difference in operating cost. Tolerating a slightly colder environment in the home can make a big saving in electric furnace bills.

Let us look at a few more items and their yearly costs. Remember the yearly cost depends on the cost per kilowatt hour of your electricity. Different geographical locations may have different rates, so costs will vary somewhat. See Table 9-3. We are sure that you know the current costs of other items as you subtract the yearly costs of some of these items from your electric bill.

## HOW TO SAVE MONEY

What can we do about saving electrical costs? The clothes dryer, like the washer, should never be operated with half or partial loads. It costs just as much to do one item as a full load, generally. Room air conditioners which have some of their radiators (or heat exchangers) clogged or positioned that they get the full amount of sunlight on hot days will cost more to operate. The dishwasher can be operated with full loads like the washer and dryer. Use of the attic or ceiling fans will help cut costs. Be sure the outside radiator (heat exchanger) of the air conditioner is free from grass, dust and dirt.

Sometimes we forget that there is a radiator underneath or behind the refrigerator. A radiator can get clogged with dirt and dust. Thus, heat is not exchanged easily from inside the freezer or refrigerator. The refrigerator runs constantly and you pay heavily. Saving on other items will depend on how you use them.

The estimated cost of electricity per killowatt hour now is about 4 cents per kilowatt hour. It is expected to increase up to 11.4 cents per kilowatt hour by the year 2000. About 16 to 18 percent of your electric bill goes for lighting the home.

Table 9-3. Yearly Cost and Wattage Use of Several Items.

| Item | Watt-Hours use/Year | Current Yearly cost |
|---|---|---|
| Clothes Dryer | 1,000,000 | 50.75 |
| Air Conditioner (room) | 800,000 | 47.50 |
| Dehumidifier | 375,000 | 20.00 |
| Dishwasher | 360,000 | 18.00 |
| Attic Fan | 300,000 | 15.00 |
| Ceiling fan | 43,000 | 3.00 |
| Freezer (15-16 cu. ft.) | 1,200,000 | 60.90 |
| Freezer Frostless | 2,000,000 | 95.00 |
| Furnace Fan | 600,000 | 33.15 |
| Microwave | 200,000 | 10.00 |
| Range & Oven | 800,000 | 40.00 |
| Refrigerator | 2,300,000 | 114.00 |
| Color TV | 300,000 | 17.00 |
| Water Heater | 4,000,000 | 215.00 |
| Fast Recovery Water heater | 4,500,000 | 275.00 |
| Air Conditioner 2,000 sq. ft. | (variable) | 250.00 |

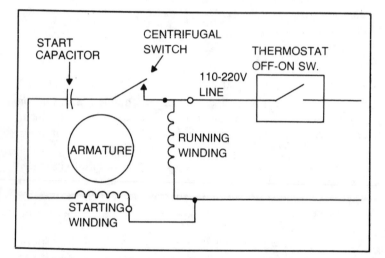

Fig. 9-1. An induction motor with capacitor start is used on many air conditioners.

## MOTOR BRANCH CIRCUITS

Some of the most costly items in the home involve the use of electric motors. One type of motor often used is the capacitor, or induction starting motor which looks, schematically, like Fig. 9-1.

### Terminals

The centrifugal switch is built in to the motor. When the armature reaches the correct speed, it flies out, opens and disconnects the starting winding and the starting capacitor. These motors are long lasting. The problem with them is often the capacitor, which may fail in some way. It is usually located outside the motor and can be replaced. Air conditioner compressor motors and the fans which circulate the air through the air conditioner radiator vents are capacitor motors, for instance. The motors may have three wires coming from their cases which go to particular terminals inside the device housing which you see in Fig. 9-2. The terminals are labeled with the letter corresponding to the wiring insulation color.

It is easy to see these terminals and their letter or code designations when they are new. As time passes, they tend to become hard to read and identify. The color of the insulation on the motor leads, however, does not tend to change. When replacing such a motor, like the fan motor in an air conditioning unit, snip off

340

Fig. 9-2. Electric terminals on a small air conditioner motor.

the leads so that about 1 inch remains on the terminal board or connecting point on the frame unit. When you put in the replacement motor, which usually has the same color-coded leads, you can easily find out where these leads connect by matching colors.

If the compressor fails due to motor failure, then you have to replace the whole unit. Make the connections to the new unit in the same manner as described in the previous paragraph. If the blower in the attic unit fails, get a replacement unit as identical as possible to it in size, make and manufacturer. If the new unit has terminals as connections, and many do have a terminal board on the motor housing covered by a small metal plate, you must carefully match the wiring from the old connections to the new connections. Take the wires off the old unit, one at a time, and put them, on the new unit. Then mount the motor in place on the frame.

If you have to mount the motor on the frame first, you may have to make yourself a drawing or diagram on paper before you take the wires off the old unit. Be sure to mark each lead that you take off the old motor so that you can positively identify it. You must know exactly where it goes when you connect it to the new motor's terminals. If there is a starting capacitor which is located separately from the motor, it, too, must be connected to the motor leads and the line. You will want to replace the capacitor if you replace the motor, even if it seems to work all right.

## Motor Nameplate

The nameplate attached to each motor offers valuable information regarding the motor capability and its wiring. Usually the nameplate will specify the motor frame and type and the following information about the motor: horsepower rating, operating frequency, phase power connections, speed, full load rated current, operating voltage(s), maximum ambient operating temperature, service factors, National Electrical Manufacturers Association design code, insulation class, protection characteristics, and, in some cases, its full load efficiency and power factor.

One example of a motor nameplate is shown in Fig. 9-3. Note how the wiring is changed for connection to a 220 volt line or a 440 (high voltage) line. This illustration is important because it shows that on some motors, even those as small as one-third horsepower air conditioner types, you may have to make jumper connections between some terminals to get the motor to run in the proper direction or to work on the proper voltage. Sometimes, just changing the jumpers in a terminal box can make the motor run in the

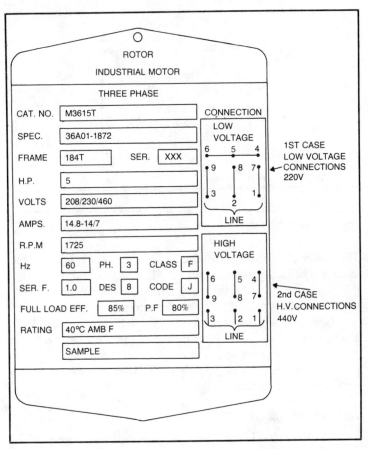

Fig. 9-3. Nameplate on a three phase, five horsepower motor.

opposite direction, and this may be desired to make everything else work as it should.

## MOTOR POWER FACTOR

The *power factor* of a motor is a measure of how well the motor uses the current running through it. There are two types of current found in a motor's windings. The *real* current is the same type that lights an incandescent lamp and the *reactive* current is not in phase with the supply pressure voltage.

The power factor is a quantity resulting from the equation:

$$\text{power factor} = \frac{\text{real current}}{\text{total current}}$$

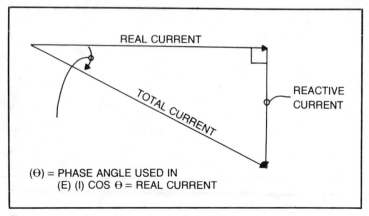

REAL CURRENT

TOTAL CURRENT

REACTIVE CURRENT

$(\Theta)$ = PHASE ANGLE USED IN
$(E)(I)\cos\Theta$ = REAL CURRENT

Fig. 9-4. Vector currents in a motor.

The higher the power factor, the better the motor is because the smaller its reactive current will be. The reactive current creates the magnetic field within the motor, while the real current does the work and causes the motor to produce power of torque.

Generally small motors are pretty inefficient because their power factors are below 70 percent. As motors get bigger, the power factors improve up to as high as 85 percent. Larger electric motors are more efficient. Lightly loaded induction motors have a very low power factor.

## ELECTRIC MOTOR LIFE

The expected life of an electric motor is about seven to 10 years, although this time can be reduced when the motor is overloaded for extended periods of operation. Overloads cause the current to be greater than the motor's rated current. The resulting heat which is generated can cause the deterioration of the motor's insulation on wires. When a motor wastes energy in the form of heat, it is an inefficient motor or is not being operated under the proper load or the proper conditions.

The efficiency of a motor is its output power divided by its input power (in the same units) multiplied by 100 percent. The efficiency of a motor may go down when it is not operated at its rated load as designed.

## SPACE HEATERS

Small electric motors are used in space heaters which have a connection to a gas line. When the gas flame heats up a plate, the

344

motor fan causes air to stream over that plate, transferring the heat to the outside air. See Fig. 9-5. You can see the gas line connection to the right. On the left is an example of a direct connection to the motor via Greenfield or BX armor cable. The cable goes to a metal box which has been fastened to a wall stud through the sheetrock wall surfacing. Inside the wall, along the stud, the "hot" electrical cable is located which then connects up through the ceiling inside the wall and runs to the circuit breaker panel located some distance away.

This example shows how a do-it-yourselfer might install a space heater for a room and have adequate protection for the electrical wiring. In the heating unit is a motor which you do not want to replace often. It is a self-lubricating type, but some types require a "squirt" of oil at least once a year into the lubrication holes provided. That occasional oiling prolongs the motor life a long time by insuring bearing lubrication.

## BELT AND PULLEY REPLACEMENT

It is wise to check belts often. If they are too loose, then the motor "slips" and does not drive efficiently. If the belt is too tight,

Fig. 9-5. The heater connects to a gas line.

then starting and stopping can cause excessive wear on motor and driven-element bearings.

If the pulley on the motor is not properly aligned with the pulley on the driven element, the belt will stretch and often jump from the pulley tracks. That stops the machinery. The pulley may become loose; many get that way because the screw which originally was tightened down into the motor or driven elment shaft gradually works back and forth until it makes a groove around the shaft. The pulley can slip quite easily around the shaft without causing any useful work to be done. You have to tighten the pulley screw(s) once in a while to be sure they stay tight.

To replace a pulley on a motor or driven element, you may have to buy a pulley remover or, as they are sometimes called, a wheel or gear puller. This little device has a bar which fits against the shaft, and a couple of arms which grab the gear or pulley. It pulls the gear or pulley off the shaft quite easily. This way is better than trying to hammer the gear or pulley off the shaft with a hammer and a cold chisel.

When you align the pulleys, be sure they are set that the motor pulley and the driven element pulley are kept in the same vertical plane. The belt should be tight but not so tight that you can't depress it about ½ inch at its center. Be sure to tighten the belt, if it is a new one, after about 25 hours of use. The new belt will stretch and expand somewhat. It needs to be readjusted; you should readjust again after another 25 hours of use if you want to do the job right. Tighten the pulleys again. Be sure you oil the motor, if it is a type which should be oiled. Oil the driven element bearings also, if that is needed. Inspect the motor once a year and adjust and maintain it as necessary.

## ELECTRIC MOTOR INSTALLATION REQUIREMENTS

The branch circuit and wires must be capable of supplying at least 125 percent of the full load current demand of the motor (Fig. 9-6). The value of the current to be considered is usually stated on the nameplate on the motor. If two or more current ratings are given, you must use the highest value rating as the basis for your calculations and selection of the branch circuit wire size. For example, if your motor has a rating of 10 amperes, full load, then you multiply this figure by 125 percent and get 12.5 amperes as the amount of current the wires must pass. You know that size 12/2 with 600 volt insulation will pass 15 amperes. That is the size cable to use to connect to this motor. You will also probably fuse the motor with a 15 ampere circuit breaker.

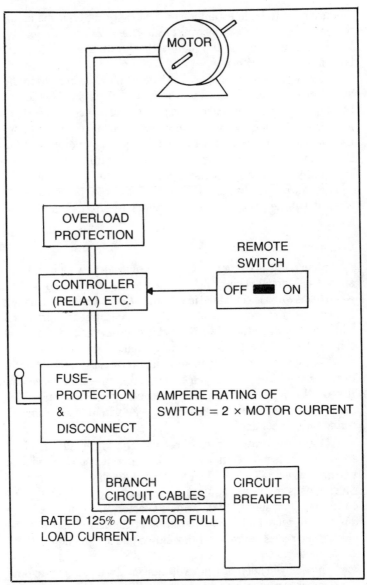

Fig. 9-6. All these items must be present for proper motor installation.

You will not normally find a circuit breaker rated at 12.50 amperes, which would be exactly the correct size to use for motor protection. The motor should have a built-in *thermal overload* device of some kind, solid state or similar, which will cause the motor circuit to be interrupted and electricity cut off *if* the motor temperature goes much above the *ambient temperature rating*. When you buy and install a motor, check to see that it has this built-in internal protection.

If you happen to be operating two or more motors from the same branch circuit, be sure the wires used will pass 125 percent of the largest current demand motor, plus the sum of currents, at full load, from all other motors which might be connected to that line. Check this even if you do not expect them to be operating at the same time. It is desirable to protect each motor with a fuse or circuit breaker of an appropriate size, even if the line is protected, and that makes good sense. In a shop of any type, home or otherwise, it will be good to have each motor run from some kind of switch.

### Controller

Figure 9-6 shows a block called a *controller*. This may be as simple as a small relay which is operated by the remote switch when that switch is contained in a thermostat. The relay is usually a sealed type unit, so you do not make any adjustment to it. If and when the relay goes bad it will probably overheat and corrode. You can tell that it is defective by a visual examination. Make a momentary "jumper" over the relay input-output terminals. If the device it operates starts to run, then you know the relay needs replacement. Remove the power through the disconnect switch and replace the relay with the same type if possible, or one with the same ampere rating on its contacts.

There are several types of relays. The instantaneous close and open type is most commonly used. It operates quickly and with a snap or even a "bang." Then there is the delay in closing type which closes a specific number of seconds or milliseconds after power has been applied to it. Some contactors and controllers use this type of relay in order to permit other things to happen in the electrical circuits before the relay closes and applies power to its motor or device.

The delay in opening relay will remain closed for a finite time after power has been cut off to its coils. When you have a controller, the prime power for the motor is always there. It is cut off or connected to the motor or device when the relay coil power is turned off or on or the relay is operated or de-energized. Cutting the power off to a relay does not necessarily mean that power is cut off to its terminals. If there is a defective unit or short through it of any kind, you can have electrical power in the unit which could kill you if you're not careful.

Some controllers are very large and complex. They insert starting resistors into a motor's windings for a specified length of time. Some have solid state control elements built into them. If you work with motors a great deal, then you should study up on the various types. Normally, in the scope of electrical wiring covered by this book, we will find only the relay types used.

## Safety Considerations

Some motors, such as fan motors, clock motors and other small ones can be frozen. They will not be damaged by not running. These need no controller as such. They are simply plugged into a wall socket. Sometimes they heat up if they are not running. Then you must disconnect them, and rather quickly.

Many hand tools, circular saws, drills, etc., which you use in your workshop have motors rated from 1 to 10 horsepower. These devices may consume quite a large amount of amperage when running. It is folly to connect one of these devices to a good branch circuit with adequate capacity through an extension cord which has a wire size good for only half of the demanded current. Don't get caught in this trap. Use an extension cord if you must, but be sure it has the amperage capacity equal to that needed by the motor at full load current.

## TYPES OF RELAYS

Relays, the electrically operated switches used in electrical control, are most useful and important. They allow the control of very high voltages and currents in a safe manner, as you do not have to handle the actual device which controls these large and dangerous electrical elements. You handle a remote switch, button or device which, in turn, activates the relay. Then the relay handles the voltages and currents.

There are many types of relays in addition to the delay in opening and delay in closing types we have already mentioned. You can find relays with multiple poles or armatures as they are called, and you can find latching and release relay combinations. This latter type, which we will find important when we discuss home security in the next chapter, will "latch" or "lockup" when a momentary current is applied to their windings. They will then hold this energized state until a release relay or coil is momentarily energized.

It is nice to be able to momentarily depress a switch or button and energize a series of motors or a motor, and have them continue running until you momentarily energize another button or switch. You can also imagine many safety arrangements which might be physically realized by having these button type contacts arranged so if any danger occurs due to the position of something, the electricity causing the feed or other action will be instantaneously cut off as the release part of the energizing relay is then activated.

### Reed Relays

There are relays which are energized by having a magnetic come close to them. These are sometimes called *reed* relays. Although they cannot pass much current through their delicate contacts, they can control larger relays called *power* relays which can control vast amounts of power. It will probably pay you to investigate the many types of relays, both the physical types and the solid state types which have no moving parts, and become familiar with them. You will then think of countless applications for many of them. Relays can also be used to make some wiring arrangements you might visualize or imagine for your home much easier and safer.

### Motor-Starting Relays

As you probably know, large motors should be started at a lower voltage than they require when running at full load and speed. To prevent an excessive voltage dip on the lines when they start and demand a very heavy current feed, it is common to energize only one-half of the motor coils either when starting or running. If the coils are in parallel, then if all were energized the current demand could be very large. If the coils are in series, then the demand will be reduced if they are fully used as the coil resistance will be greater. In any event it is desirable to use the arrangement which will demand the least current and cause the smallest voltage drop over the lines when starting the motor. The relay makes this possible by switching out or switching in coils as appropriate to whether the coils are in series or parallel at starting.

After the motor builds up speed (and counter electromotive force,) then this back electromotive force can be used to trigger a release relay. The relay causes the rest of the parallel coils to be placed in the circuit, or causes the series coils to be reduced to one-half their starting value. The motor can run fast and produce the required torques for handling the loads. When the motor builds up back electromotive force by running at near full speed, it demands less current and will not cause the voltage dip or drop that it might cause at starting without the use of relay starting controls.

## SHADED POLE AND SPLIT-PHASE MOTORS

The main types of motors found in industrial applications are single-phase single-voltage types, dual-phase single-voltage types, and three-phase single voltage types. There are also three-phase dual-voltage types and three-phase two-speed types of motors. These may be shaded pole, split-phase, induction-repulsion, universal ac-dc, three-phase wound motors and three-phase synchronous motors.

The *shaded pole* motor is used in small appliances. Usually it ranges from 1/100 to 1/8 horsepower in size. Inside there is a coil called an armature, and there is a thick ring of copper, called a shading coil, which is physically fastened to a pole piece. This shading coil causes such currents that the armature drags around and rotates causing its input-output shaft to rotate. If you want to reverse the direction of rotation of this kind of motor, simply remove the armature, turn it around and put it back together again. It will then run in the opposite direction.

Split-phase motors are small motors ranging in size from 1/20 to 1/12 horsepower. You will find them doing relatively small but important jobs in the home. They can be found with washing machines, exhaust fans and garbage disposals. The motors operate with either 115 volts ac or on 220-230 volts ac. When connected to a single-phase line, meaning that they are plugged into a normal wall receptacle, the motors will run and perform the task assigned to them. The split winding makes it possible for them to be self starting and to run on a single-phase current.

## INDUCTION-REPULSION AND UNIVERSAL MOTORS

The *induction-repulsion* motor uses less starting current than any other type. It also has a very large starting torque. For these reasons, it is normally used for pretty large loads. This motor does not have a wound armature. It will have an iron armature which has copper conductors inserted into the iron and connected at each end. When energy is applied to the motor windings, this causes a changing magnetic field. A magnetic force develops in the armature which is repelled by the magnetic field developed in the fixed windings. The armature, trying to move away, rotates. The induction-repulsion motor is used on many heating-cooling units.

A universal motor is one which has a commutator and brushes. It will run on either alternating or direct current. The armature arrangement will so control the direction and position of the magnetic fields that any applied voltage of the proper value will make this type of motor run. As with any motor which has an armature, the maintenance is higher because the brushes and the armature wear down. There will be sparking at the armature when this happens or when the brushes are not properly positioned around the armature circumference.

## THREE-PHASE MOTORS

Three-phase motors having a wound rotor and the three-phase synchronous motors both operate on three-phase current. You must supply them with this kind of current from the wall plug, switch or controller. Wound rotor types are used where the motor speed is adjustable and you want to vary it. The amount of current in the rotor can govern the speed. There are three windings on the rotor which are positioned 120 degrees apart. Each is connected to a continuous type slip ring. A slip ring is not a *commutator*. A commutator has many segments while a *slip ring* is a continuous metallic band of copper around and insulated from the armature metal frame. The slip ring type is always an ac motor while the commutator type is a universal motor described previously. External resistances in a controller box can be used to control the current and voltage to the three-phase motor armature and thus control its speed.

Three-phase synchronous motors run at the same speed all the time, no matter what their load may be. Since there is no speed variation, you will not find, usually, the armature with any windings upon it or through its core. A rotating magnetic field in the *stator* or frame windings of the motor will cause currents to be induced in the armature copper bars; the magnetic field produced there will try to follow the stator magnetic field in rotation. The rotating of the stator field is governed by the frequency of the applied voltage-current. At 60 hertz, for example, the armature will try to reach the speed of 3600 revolutions per minute. It does, but not in exact phase relationship with the stator rotating field. The armature slips behind in space degrees according to the motor load.

## VISIT A MOTOR SHOP

If you are into motors, visit your nearest motor repair shop and learn from them. Observe motors being taken apart. Try to put one together. Learn how to test and repair motors. Be sure you know the full details of installing new motors for any heating-cooling system you might have in your home. It can save you some money. Replacement of motors in washing machines, disposals or on small fans probably will not be worth the effort. It's better to get new units. When visiting a motor shop, ask how you connect the wiring for forward and reverse rotation of the motor.

## RUNNING THE WIRING

At first thought it might seem that all you do is to run a nonmetallic cable to the motor and connect it. That is not quite true. When wiring a motor, you must realize that is a mechanically moving device. The motor will create vibration of some amount and will jerk frames it may be mounted to as it starts and stops. The larger the motor, the greater this reaction torque can become.

When we run wires for motor connections, they must be the proper size, insulated for the proper voltage (normally at least 600 volts), and have insulation capable of withstanding the environment in which they will be placed. All these factors must be considered in the selection of wiring to be run from the circuit breaker and switching device or controller to the motor itself.

In a majority of cases you will find that rigid conduit is used to make the channel for the wires to run in. This conduit goes from the circuit breaker panel to the switch box and to a box near the motor, if the motor itself doesn't have an electrical box as part of its frame construction. If the motor does have such a box, then the conduit may run directly to that box. The conduit is securely fastened to the metal box with good strong fittings designed for the purpose. Then the wires are run inside, using a fishline or similar type of pulling device to pull the wires through the conduit.

Where there is no danger of dampness or water, you will find that BX cable may be used. This is a flexible armored conduit with wires inside it. When connected to boxes with the proper fittings designed for it, it makes a very strong channel for its wiring. You can buy it with the kind of wire insulation you need for that particular job. The flexible cable permits a certain amount of motor vibration and "play" in the motor mounting without causing trouble.

### Air Conditioner and Refrigeration Arrangement

In some cases, like an air conditioner of the evaporative type or a blower in a refrigeration unit, we have inspected arrangements where the incoming wires are run through rigid conduit to a box which is mounted inside the enclosure of the motor-operated device. This box is then provided with a proper receptacle outlet so that a short flexible cable from the motor with a plug on its end can be pushed into the receptacle socket easily to obtain power. This arrangement makes it possible to quickly disconnect the motor when you have to work on it. The non-metallic sheathed cable from the motor is usually very short, 1 or 2 feet at most. It has insulation designed not to fray, crack, rot or permit water entry. Even so, plan to examine the cable of ten, say every four to six months.

## Shop Arrangement

In a shop arrangement, you might find it worthwhile to use a flexible cable from the motor to some general outlet box mounted next to a good size disconnect switch. This switch is fused and mounted near the multiple outlet box. You might use plugs as mentioned, or you might have fittings and BX flexible conduit with wires in it. Or use Greenfield, which is a flexible armored conduit without wires in it. You run your own wires inside it.

The armored conduit is recommended if there is a possiblity of having anything hit that cable and wiring to cause damage. Sometimes we can hit an electrical cable and jerk the connections at the ends. But the device still seems to work all right. What we don't know and realize, is that the jerk caused a loosening of connections somewhere. If you use armoured conduit with the proper fittings and make them tight and secure, it takes the jerk impact and usually won't suffer from the accident.

Running the wires in conduit also keeps dirt, grime and other chemicals from getting on the wiring which can cause real trouble. Oil can accumulate on the wiring easily, and I've often wondered by what magical means this can always occur! Even with strict preventive measures, oil, grime, dirt, metallic chips and other things appear on the wiring.

## Coating The Terminal

In the terminal box which is on the motor, be sure to make good solid connections. Do not strip the wire insulation back so far that bare wires are exposed. Be sure the wire makes one complete revolution around the terminal screw or bolt with the insulation up snug to the screw or bolt at the starting point. Tighten your nut, washer and lock washer securely so that you make a good electrical contact. If possible, coat the whole terminal with a good grade vasoline grease, nail polish or even an enamel paint to keep the terminal from rusting. If you have a terminal where spade lugs or hole lugs are necessary, crimp these to the wire ends good and tight. Then secure them in place by their screws or bolts and again coat the terminal.

If you use a color substance to coat the terminal, be careful. Sometimes the manufacturer will color code the terminals using a drop of white, blue, red or other colored paint, polish or varnish. Perhaps a little vasoline or heavy grease on the terminal might be better than any kind of paint or nail polish. If the color code is not perfectly clear, you won't know which wire to reconnect to what point.

Vasoline or grease is not appropriate if the motor is in a hot location. The important thing is to inspect the connections often.

We do not necessarily use a sheathed plastic cable, as used in house wiring, to connect to motors. We might use this cable to run to a plastic box in the vicinity of the motor installation, and then go from there with conduit or armored flexible cable or conduit. But we do not recommend plastic boxes or plastic sheathed cable to connect to the motor directly unless it comes in a proper conduit.

## RELAY TROUBLES

If you have a relay controlling your motor, it can give you some trouble. Perhaps you start your motor and it won't shut off. The relay points may have been welded together. The armature may stick due to dirt. It might also stick because wear during operation has changed the clearance distance between the relay armature and its pole piece. Sometimes the armature return spring will break and the armature then cannot pull out to stop the motor.

If you push the on button of the motor controller and the motor does not run, your relay may have a burned out coil. A contact to the coil of the relay may be loosened or corroded so no electrical contact to the coil results. Possibly the relay armature-pole piece gap has been widened somehow due to vibration and wear so that the armature cannot pull in as it should to make a proper electrical contact to run the motor. Then, too, there may be occasions when dirt or grime gets on the relay contacts.

Inspect the relay occasionally and turn the power off to its contacts. Clean the contacts by burnishing them with a very fine file. Do not file them harshly or use rough sandpaper to clean them as this wears away the silver contact metal and makes the contact of the points uneven. Check the spring and the pole piece-armature clearance by pushing down on the armature. The contacts should be made without having the armature touch the pole piece. Figure 9-7 shows the critical parts of a relay.

There is another type of relay associated with an electric motor, which you don't see very often but you hear it operate when the motor starts. This is a *centrifugal force* relay device built into the motor. It closes a set of contacts to a starting winding of the motor so that it will start easily under loads. Just as soon as the motor builds up some speed, the little weights at the ends of some small arms fly out or open up away from their resting position. They open the circuit to the starting windings of the motor and leave the running windings connected. You can hear this movement as a bang, a click, or other similar sound when the motor starts running. If the centrifugal force relay doesn't operate correctly, you can burn out the starting winding and the motor will not run. We seldom have trouble with this kind of relay unless the motor is operated in a very bad environment. Then you must check them every four to six months.

Clean, check and adjust if necessary a centrifugal force relay. Sometimes it is not practical to repair such a relay. It is better then to get a new part or to buy a new motor if parts are not available and the starting relay is giving trouble.

## MOTOR SPEED CONTROLS

With some larger motors you can change the speed by varying a resistance in series with the armature winding. A good controller element will make this possible when it is wired into the motor circuit. With some motors a low speed winding and a high speed winding are available. Either of these windings can be used indefinitely to run the motor at either its high or low speed or operation.

You need a single-pole, double-throw type of switch to control the speed. There will be three lines to the motor, a common line and two "hot" lines. One "hot" line will go to the low speed winding and the other to the high speed winding. Figure 9-8 shows an arrangement and wiring for this type of motor.

These motors are quite common in air conditioning and heating units. The wall switch shown is the single pole, double throw type needed to operate this motor. Note that its off position is in the center. Usually the motors do not draw over 15 amperes, as the 12/2 cable is large enough. If the motor does draw over 15 amperes, then use a 10/2 or other suitable size.

Another way of controlling the speed of a small electric motor is shown in Fig. 9-9. This unit uses a triac to control the

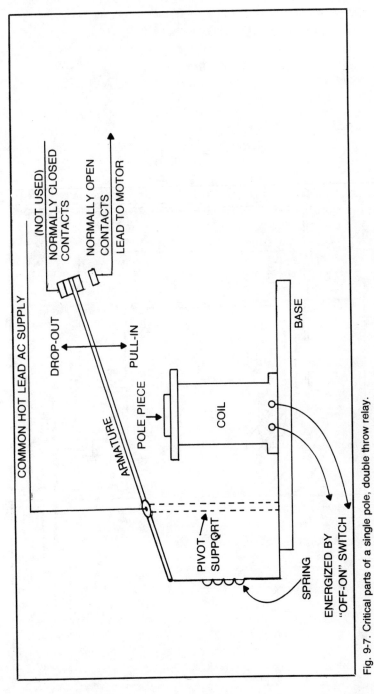

Fig. 9-7. Critical parts of a single pole, double throw relay.

COMMON HOT LEAD AC SUPPLY

(NOT USED)
NORMALLY CLOSED CONTACTS

NORMALLY OPEN CONTACTS

LEAD TO MOTOR

DROP-OUT

PULL-IN

ARMATURE

POLE PIECE

COIL

BASE

PIVOT SUPPORT

SPRING

ENERGIZED BY "OFF-ON" SWITCH

357

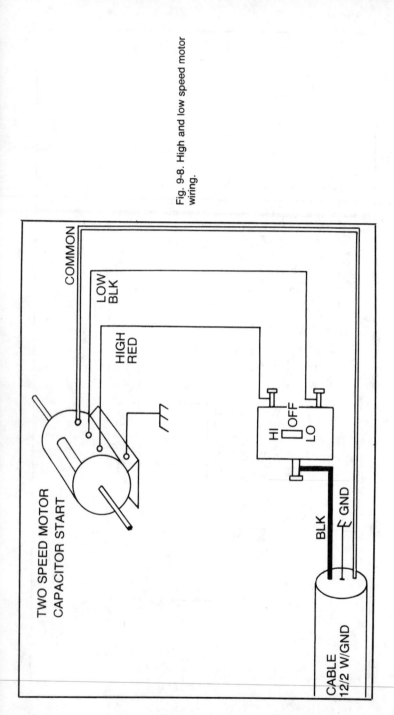

Fig. 9-8. High and low speed motor wiring.

TWO SPEED MOTOR
CAPACITOR START

COMMON

LOW
BLK

HIGH
RED

HI
OFF
LO

BLK

GND

CABLE
12/2 W/GND

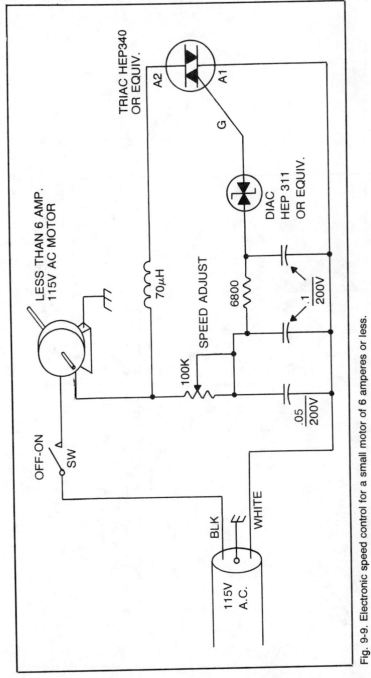

Fig. 9-9. Electronic speed control for a small motor of 6 amperes or less.

359

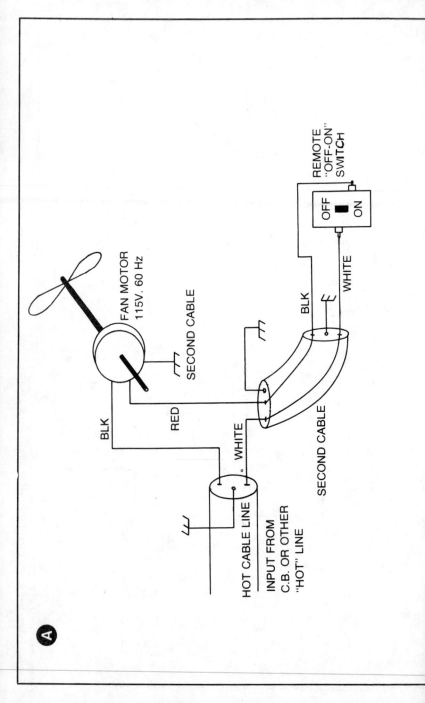

FAN MOTOR 115V, 60 Hz

SECOND CABLE

REMOTE "OFF-ON" SWITCH

OFF

ON

WHITE

BLK

BLK

RED

WHITE

SECOND CABLE

HOT CABLE LINE

INPUT FROM C.B. OR OTHER "HOT" LINE

A

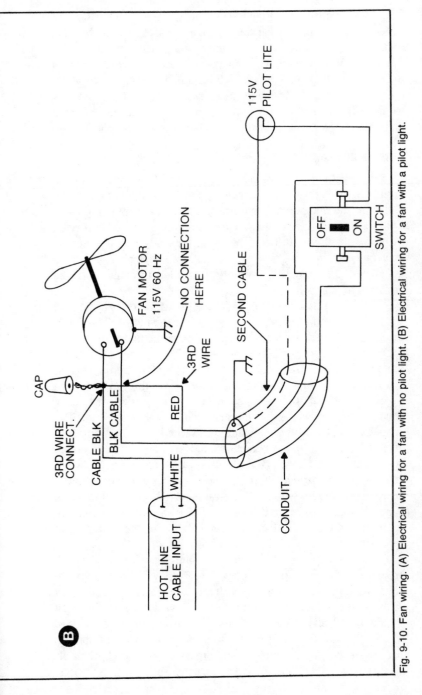

Fig. 9-10. Fan wiring. (A) Electrical wiring for a fan with no pilot light. (B) Electrical wiring for a fan with a pilot light.

Labels in figure:

PILOT LITE
115V
OFF
ON
SWITCH
SECOND CABLE
FAN MOTOR
115V 60 Hz
NO CONNECTION HERE
3RD WIRE
CAP
3RD WIRE CONNECT.
CABLE BLK
BLK CABLE
RED
WHITE
CONDUIT
HOT LINE CABLE INPUT
B

flow of current to one side of the motor. There is a potentiometer speed control to permit varying of the amount of current the triac will pass to the motor. This circuit may also have use as a light dimmer type of circuit wherein you replace the motor with an incandescent light or lights which do not, in total, draw over 6 amperes of current. That could mean two 250 watt bulbs or as many as six 100 watt bulbs, although you might just use one 100 watt bulb. Parts are available from such places as Radio Shack.

## PUMP MOTORS

We must not overlook pumping motors which come in a wide variety of sizes for many different applications. Pump motors are built into the pumping mechanism in the usual small pump case. When larger pumps are used, the motor may be separate and driven through a shaft of some kind. The motor is usually synchronous and may range from a fraction to several horsepower. In irrigation situations, the motor is remotely controlled through switching circuits and controllers.

When putting in a pumping device, the first step is to find out what circuit requirements are specified for the motor. This information can be determined by examination of the motor nameplate data. The circuit should be planned and laid out or roughed in. Then the motor and pump assembly should be mounted and the conduit and wiring roughed in. That means getting everything mounted, all wires in place and other tasks completed except making the actual connections to power and to motors, switches or controllers. The pressure switch, if used, should be attached and its wiring connected. Then the motor terminals should be located and connected, being careful that all watertight fittings, plates and boxes are water proof when you have made the connections. Grease the terminals or waterproof them with an electrical insulating spray. Rotate the pumping shaft to be sure it isn't locked for any reason. Then turn the power on and check for proper rotation of the motor as to approximate speed and direction of rotation.

If the pump and motor are one unit, then the motor will probably turn in the proper direction. If they are two separate units, then you might get the wiring to the motor reversed so it turns in the wrong direction. Be sure everything works as it should. Check all lines and cables and be sure you have no exposed wiring. Use rigid conduit or flexible armored conduit as appropriate. Rigid conduit is best for complete waterproofing. In some air conditioning units the motor and pump will be submerged in water.

Be very careful that you do not get any loose wires which are not designed for underwater use placed that they are or can get in the water.

## EXHAUST FANS

We have already discussed the installation of circulating fans inside the house or apartment. Now we consider the larger exhaust fans which might be installed in an attic so they will vent the hot air which forms there and reduce temperatures inside the home by keeping the attic cooler. These are relatively low cost items. They may be obtained in many sizes, from 18 inch to 48 inch size blades. Of course, the size of the motors must be compatible with the size of the blades and the size of the loading.

The motors are normally a permanently lubricated type and require little attention. Tpe only problems are the installation mechanics and the electrical connections. The mechanics must be such that any vibration from the fan due to uneven blade rotation or blade weight won't cause the fan to get loose and damage something, or be prevented from running. The electrical part of the installation should provide for a remote switch inside the house with a pilot light to show you the fan is on or off. There could be an automatic thermostat control for the fan so it will energize when the attic temperature reaches a predetermined level and continue to run until that temperature is reduced to some lower level. Sometimes the temperature can't be reduced and the fan just keeps running in very hot climates.

You can run a 12/2 cable from a box near where the fan will be down into the house through a roof or room hole. You will then seal and attach a switch to the ends of this cable as shown in Fig. 9-10A. If you need a pilot light, which can be small in size like a Christmas tree light, use a three-wire cable with ground (making four wires in the cable) and connect it as shown in Fig. 9-10B. If you use a thermostat, the instructions with it will tell you how to connect it to your fan.

## SUMMARY

So we close this important chapter on motors and electrical costs. There is a lot more to know about motors, especially, but you have most of the basic facts. With motors, as with most electrical wiring, a lot is common sense. Use common sense in making good connections, turning power off when working with electrical wiring and connections and making good mechanical installations.

# Home Security Systems 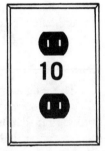 10

You should have a home security system which is electrically or electronically operated. The first step you should take to secure your home, if it is guarded by an electrically or electronically operated device, is to protect that all important circuit breaker or fuse panel. If it is possible for an intruder to easily open up that panel and throw the master circuit breaker or switch to disconnect all power inside the home, then none of the devices which we shall discuss will be suitable. Of course, we do not rule out those which change over to battery operation in case of power disconnection or failure. These might still work and give some kind of protection.

## PROTECTING THE CIRCUIT BREAKER OR FUSE PANEL

How do you protect the circuit breaker or fuse panel? Put a padlock on it. Most panels are designed so that when the cover is closed, there is a place for a padlock which will prevent easy opening of the panel cover and access to the interior. Check your panel right now. If it is inside the home, it probably doesn't need a padlock, especially if it is in a hide-away closet. If the panel is in a garage, shop or outside the home, then you do need a padlock for that cover.

When you put the padlock on the cover, get a good one as cheap padlocks won't stand up well under weathering. They may rust out and then you can't operate them. Make it a general practice to oil the padlock every six months at least. A few drops of oil around the loop openings and into the key opening will do wonders. Put the key away carefully with a large tag on it saying what it is so that you won't lose it.

We will now develop some security concepts based on the assumption that you will have primary electrical power and your source is secure, at least from normal interruption. We say normal because if you have attended any police lectures lately, you have learned that it is almost impossible to keep an experienced thief out of your home if he decides to break in. You just make it difficult for him, and he might change his mind about doing it.

## TIMING LIGHTS

These lights are available in stores. They are timing switches into which you can plug a lamp. You can set them to go on and off at predetermined times. Some of them will come on and go off at random times. It has been suggested that a pattern of lighting, using such timers, be used in the home which is to be vacated for a period of time. Set one timer in a room so a light comes on at, say, 5 p.m. and goes off at 8 p.m. At the same time set another timer in a second room so that it will turn on its lamp light at, say, 8 p.m. and turn it off at 9 p.m. Another timer and lamp can be operated in a third room from 9 to 10 p.m. Thus, you give the appearance of having various rooms lighted and dark just as they would be when someone is occupying the home. Also, you get more protection from using several timers instead of just one. If one fails, the others still work.

It is possible to buy a commerical timer which you can install on a separate branch circuit from the circuit breaker panel. It can be locked if it is installed outside and will control lights on the outside of the home. Separate branch circuits from this type of timer can be run to lights and thus control them. If you find this idea attractive, then talk to your local, professional electrician or the electric company. They will be glad to guide you as to what you'll need. Thieves don't like to work in the light. Lighting up the outside of a residence can discourage them.

When using timing lights, set them the day before you leave the home and check them out. We have found many timers which don't operate, stick or bind. Others tend to run slow or fast so that the timing is all fouled up. It's always a good idea to check them to see if they do turn the lights on and off at the set times. Check these units by feeling them with your hand after several hours operation to see if they are uncomfortably hot. They might have some defect which could cause fire. Don't take for granted that these timers will work properly.

## BASIC HOME SECURITY LIGHT AND WARNING CIRCUITS

There are two basic circuits which are used in home security electrical circuits. These are the parallel circuit, shown in Fig. 10-1A, and the series circuit which is depicted in Fig. 10-1B. The parallel circuit can be used for multiple doorbell ringing purposes if you have one switch at the front door and one at the back door. In that case *you use a step down transformer,* and the voltage on the line is only 6 to 10 volts ac.

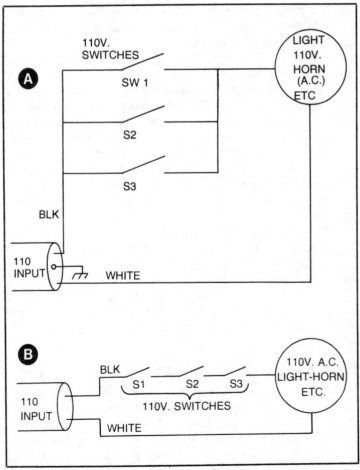

Fig. 10-1. (A) Basic parallel security circuit. (B) Basic series security circuit.

As shown in Fig. 10-1, we are considering the lighting of a 110 volt light, or lights, from different places using 110 volt type switches. Each of these lights may have its own wall switch already in the the room where it is located. Sometimes, though, you may want to turn on that light from a long distance away or from upstairs. You might want to turn on a garage light from inside the house as well as being able to turn it on from inside the garage. The parallel switch circuit will enable you to do this. Note that if you turn the light on from any switch, no other switch can turn it off. This is a bad feature of the parallel circuit arrangement. If you need to be able to turn the lights off from any place the switch is located, then we refer you back to the wiring of the two-way and three-way switches we discussed earlier regarding hall and stairway lighting. Using "traveler" wiring, that arrangement can turn off and on the lights from different positions.

In Figs. 10-1B you have the basic security wiring system used continuously around the home from each doorway to each window, etc., until the whole set of possible entry locations has been wired. This kind of series circuit works on the premises that the circuit will remain energized as long as no one forces an opening into the home through a doorway or window. If someone does and the switches (magnetic types) separate, one part from another, the circuit will be broken. That will cause a relay or some device to "drop out" and make an electrical contact which, in turn, will cause an alarm to sound or lights to come on. Figure 10-2 shows an alarm and magnetic switches.

As shown, the circuit will simply cause a light to go out if any switch is opened. Thus, in a protective role each switch is normally closed. Any interruption to the circuit will cause a switch to open, shutting off power in the circuit and then through the proper auxiliary alarm equipment, causing it to sound when the power to its control element is cut off. Naturally there will be other power to the alarm unit which is not dependent on this series circuit, except for control.

## ADDING A REMOTE SWITCH TO AN EXISTING LIGHT SWITCH

Now let us consider the parallel circuit and its possible use in a security situation. Suppose that you have a room downstairs where there is a light switch and that you sleep upstairs, as many do. It would be nice, if you hear a noise at night, to turn on a switch and cause that downstairs to come on, brightly, which would cause an intruder to leave the premises immediately.

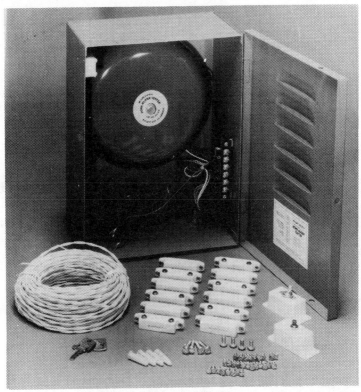

Fig. 10-2. Magnetic switches and an alarm bell.

We can add this little remote capability very easily by paralleling the existing light switch with a length of sheathed cable, run nicely, carefully and inconspicuously along a wall, or up through siding and over the attic's interior to the new switch location. It will terminate in a box set into the sheetrock or whatever. In Fig. 10-3 we show the electrical connections to be made. All you do is simply parallel the existing switch by connecting the black and white wires from the cable to each side of the switch. Then run them to the new switch located somewhere else.

The new extension cable can be as long as desired. If the old switch controls a receptacle socket, you can control a lamp plugged into that receptacle with both the new and old switches.

You can control the lights in any room, basement or garage from a panel of these remote switches. They will each be wired parallel to some existing switch. That would give you the capability to turn on lights anywhere, in any sequence, throughout the home grounds or garage. It just takes some added sheathed cable and some new switches.

Be aware, again, that if you turn on any light from the remote position, you cannot turn it off from the old switch position. The only way you can get that light off is to be sure all parallel switches are off.

Expanding still further on this idea, we next suppose that you have run parallel cables to a number of switches all located in the same place. It is easy to provide an electrically operated multiple-switch timing device to operate those remote switches by again paralleling the remote switch with a timer switch. You can set the timer so it will switch various lights on and off at various times throughout the house. You might not want to have remote switches at all, but just a timer which will operate when nobody is in the home. That is easily installed according to the instructions which come with the units when you have the circuits in place and connected to the remote switches that we have described.

Perhaps you can connect all "common" leads together and run these through another larger switch, which can serve as a "master" control switch for the remote switches. When it is off , no remote switch will operate. The whole electrical system in the house will be operating as it would normally. When that switch is on, then all remote switches are connected and you have the double "turn on" capability. See Fig. 10-4. You must have the common wire from each branch circuit isolated so you know it is the common wire; otherwise, you might might create a short circuit connecting wires together.

In Fig. 10-4 you see how the hot black line from the incoming cable is connected to each black wire of the switch cable. The white wire from the incoming "hot" cable goes to each lamp wire directly and there only. The second lamp fixture wire goes to each white wire of the remote cable and, from there, at the other end of the cable to one switch terminal of each existing remote switch. Meantime, the first remote cable has its black wire connected to the common black wire at the top, and then through the master disconnect to the opposite or unwired terminal of each remote switch. The black wire from each remote cable goes to a terminal of the normal, existing switch. If you now follow the continuity for each switch pair, you see that turning them will light the light if the master disconnect switch is turned on also.

This concept can be extended for as many lights or devices to be operated by switches as you desire. It just takes a lot of cable

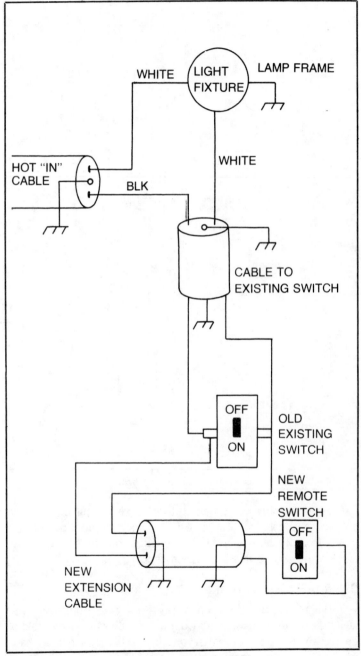

Fig. 10-3. Adding a remote switch to an existing light.

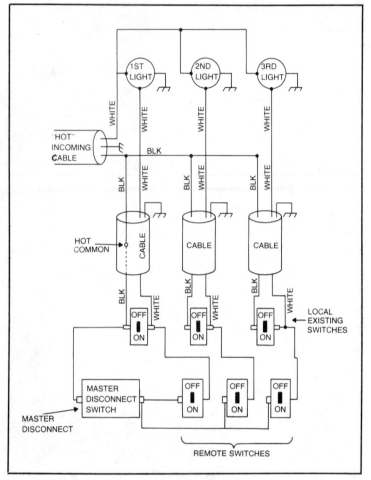

Fig. 10-4. A master disconnect for remote switches.

and some planning and checking to be sure you have the wires
properly connected so as not to have any short circuits. Be sure the
power from the master circuit breaker in the circuit breaker panel
is off while you do this wiring and checking.

## USING THE TIME DELAY AND LATCH RELAYS

When you consider secondary or tertiary control from a given
switch, you are thinking about the use of relays or some kind of
device which is activated when you close that first switch. We have
mentioned delay relays in motor control. They can also be used in

372

security control. You can buy these relays with adjustable time delays from a few seconds up to many minutes. When you close a circuit which energizes them just for a second, they will remain energized for a long time thereafter until they finally "drop out" and de-energize the circuit by their own action. That helps to make life simpler because we don't have to worry about turning things off.

A nice arrangement for this kind of circuit is the doorbell. Connect such a relay to the bell circuit in parallel with the bell or chimes. When someone comes to the door and rings the bell it can cause, through this time delay circuit, a light to come on at the same time the chimes are rung. The light will stay on for several

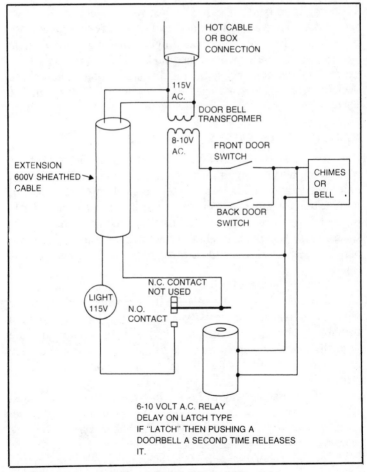

Fig. 10-5. You can operate a light with the doorbell switch.

minutes until you can (by another parallel switch) turn it on for as long as you need it.

There may be instances where you do not want a time delay relay to operate and shut off automatically. You might want a relay which will "latch" in the on position and stay in that position until the doorbell is pushed again, at which time the light will go off. In that case you will use the *latching relay*, available from electronics stores or relay manufacturers. Use it in the circuit shown in Fig. 10-5.

Another warning alarm unit is the mat switch, type CVP 1723. You can place the mat at the doorway and wire it to a light or bell. If anyone steps on the the mat that light will burn or the bell will ring. There is also a type 180-S flex switch which might be used in a screen door hinge area. If any one opens that screen, or any other door, that switch will be activated by the bending of the flexible covering on it. These switches are rated at 117 volts ac and can be used with the regular house lines and branch wiring. Obtain these switches from *Allied Electronics*, 401 East 8th St., Fort Worth, TX 76102.

## MAGNETICALLY OPERATED REED SWITCHES

Many home security systems available on the market for installation by you have the reed switch, magnetically operated type of device with a power source and an alarm—either a bell, horn or other audible device—to let you know that an intruder is present. The circuit used in these systems is the series circuit. Each switch shows the sealed magnetic units consisting of two parts: one section houses the switch and the second unit contains the magnet.

When the magnet is close to the switch unit, the switch will close and make the circuit complete. If the magnet is moved away, which would happen if a window were raised or a door opened, then the circuit is broken. This can cause an alarm to sound through the normally closed contacts (or the de-energized contacts) of a relay in the circuit wiring. Figure 10-6 gives an example of this type of circuit.

You need some way to insure that the system is activated when you turn it on. Control the electricity to the system through an off-on switch as shown in Fig. 10-5. Use one off-on switch in the main branch circuit power lead to the transformer.

Testing the system without ringing the bell or sounding the horn may require a small light positioned at A in Fig. 10-6. This

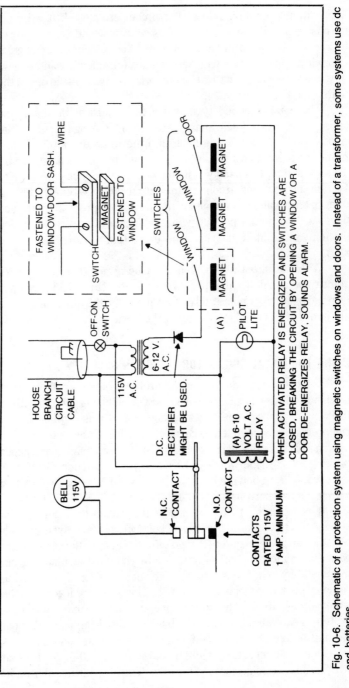

Fig. 10-6. Schematic of a protection system using magnetic switches on windows and doors. Instead of a transformer, some systems use dc and batteries.

375

light will not come on unless all switches are closed. It will then burn brightly as long as the switches remain closed and the power is applied to the protection system. You can buy rectangular ceramic or plastic units with screw holes for mounting and screws or wires for connection to the switch which is built inside one of the units (see the insert in Fig. 10-6).

There is a second type which is cylindrical in shape and appearance, which you can hardly detect by looking at its installation (Fig. 10-7). It goes in flush into the wood or metal window frame and sash; its wires can easily be hidden so they can't be seen. This second type is used when you want no evidence of a security system and you need to connect the output of the system (at the bell or horn terminals) to a telephone line through appropriate interface devices. You do not put 115 volts on such a line so that a police station can be alerted. This type of system is called the silent alarm.

With the silent alarm, you must be aware that it may take some time for a police unit to get there. In that time, a thief may get into and out of your home and be long gone. On the other hand, if no alarm or light is audible or visible, the intruder may take his time and be caught.

## USING THE METAL CONDUCTOR RIBBON IN SECURITY SYSTEMS

For a security system you will probably obtain a roll of metallic ribbon which has glue on one side. The ribbon is used to outline the window glass so that any breakage of that glass will break the ribbon, which is conducting current. It will then act like an open switch and cause the alarm to sound. The ribbon usually has a paper backing on one side which you have to peel off. Then the glue is exposed. The ribbon sticks to a cleaned window very nicely when pressed down hard into position.

It is essential that the ends of the ribbon be connected in series with the other switches and conductors of the system, so it becomes an integral part of the system. If that metallic ribbon must pass over a metallic part of the window you must insulate it from that metal by using some tape or thick paper. There is no need to cover or provide any insulation for the ribbon where it touches the window glass itself. The voltage present on this ribbon when the system is activated is only about 6 to 12 volts dc or ac, so it is not harmful to any person touching the ribbon at that condition.

One possible arrangement is shown in Fig. 10-8. Notice that two flexible wires go across the hinge. They are insulated from it,

Fig. 10-7. Cylindrical switches, horn and key lock.

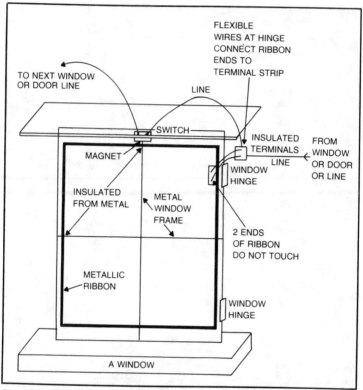

Fig. 10-8. Place the metallic ribbon on the window properly.

of course, to connect the ribbon ends to the fixed terminal. When the window opens, it breaks the magnetic switch circuit. If the window glass is broken, the ribbon will be cut or broken and the alarm will sound.

We have indicated a window with a metal frame and sash to show how you must insulate the ribbon from the window's metal parts. If you have a window with a wood frame and sash, the installation is much easier as you do not have to worry about this short circuit possibility. Also, if the window goes up and down, then you need a different method of connecting the ends of the ribbon to the fixed terminal so the wires won't be broken when the window is raised or lowered. We leave that mental exercise to you and your ingenuity. Sometimes you will be unable to use the ribbon due to connection difficulties.

Most security systems can't absolutely keep intruders out, if they want to enter. However, they usually will not attempt to gain entry if there is any doubt at all as to whether you have a working security system. It is necessary to get some signs which say, that your home is protected by a security alarm. Be sure these are pasted on windows and doors so they are visible. Of course, you want these with actual wired systems also. Make sure that these signs are easily seen from outside by anyone passing by or coming to the door. Figure 10-9 shows one type of alarm unit which can hang on a door.

## FIELD DISTURBANCE ALARM SYSTEMS

Now we will examine some more methods of creating alarms when motion is evident (electronically) to a home sentry device. In

Fig. 10-9. A door alarm unit.

Fig. 10-10 you see the Safehouse RF field disturbance alarm system available through Radio Shack. It can protect an area of up to 50 feet in length and a volume of up to 5,000 cubic feet. The unit radiates a beam of invisible radio frequency energy which, when disturbed, causes an alarm to sound.

This unit is said to be secure from false alarms due to wind, traffic noises, air conditioner operation, telephones or winds. It will operate only when someone walks through its field of sensitivity. You set the unit by pointing its grilled surface toward the area it is to protect. You turn it on and then have some 30 seconds to get out of the room so it won't set off the alarm when you leave. When returning, you will have 20 seconds to turn it off before it operates.

There are many types of field disturbance units available on the market. You will want to consider the unit's coverage in square feet, its safety backup capability in case electricity fails, and how you can test it to be sure it is actually covering the space it is supposed to cover.

There is a built-in test light which will show you that the unit is operating as it should be. If you walk through the protected area, the light will come on. You can see what the dimensions are of the sensitive area and can locate any blind spots or problem areas. Some systems may have an external detector, which can be carried around in your hand as you walk, to detect the strength of the radiation field.

## ULTRASONIC ALARM SYSTEMS

Operating in about the same manner as the radio frequency field radiation units are the *sonic* intruder alarm systems. These operate with a frequency just above that which can be heard by the human ear—about 25,000 hertz. The unit sends out a beam of sound waves which will be reflected back to it when anything moves in that beam or field. The returning echoes will cause the siren or bell to operate.

The Radio Shack Safehouse deluxe ultrasonic home alarm is one of these type units. It looks very much like that shown in Fig. 10-10, and has the same type of delay features. An intruder will be in trouble unless he knows exactly where the system is and how and where to turn it off. Don't let anyone know about any security arrangements you might make for your residence.

The ultrasonic unit has a variable area coverage switch for areas of about 10 by 15 feet, 13 by 20 feet or 20 by 30 feet. See the sensitivity for the area you want to protect.

Fig. 10-10. The Safehouse RF field disturbance alarm system.

There is not much wiring necessary for these units. They have a battery supply built into them which is charged when the unit is plugged into the wall receptacle. It keeps charging at a low rate so that if the electrical power fails for any reason, the battery will still operate the unit. Just be sure that you do plug the unit into a wall socket and that your plug makes good contact with the wall socket. Check to see that the unit does not become unplugged.

## SECURITY SYSTEM WIRES

Regarding security systems installations, a telephone extension cable, which consists of # 22 to #24 wires, connects the various magnetic switches and devices in such a system. The wires can be obtained in a pair in a cable which can then be easily stapled to baseboard or window frames or in wall corners. The color of the cable should blend into the room coloring so nicely that the wiring is not noticed normally. If you can run the wiring into the roof attic space and then down inside the wall, then it won't be seen at all.

## TRICKLE CHARGE BATTERY

Every good system must have a battery which operates it. That battery must be on trickle charge all the time and be a type that trickle charging will not damage. Normally a 12 volt battery is used in most systems. The main, or control, unit which houses the battery, charger and switch for off and on operation of the system is connected to the house wiring with a permanent type connection in a box or through sheathed branch circuit cables running to the circuit breaker panel. You don't want this unit to be unplugged by anyone. Also, you do not want anyone to just turn the security system off and on at will using a normal type electrical switch. You want a key-operated switch to operate the unit, and you do not want anyone to get that key or a duplicate of it.

## SIRENS

A siren like in Fig. 10-11 is probably the best type of alarm. Two such units might be used. One can be inside the house to scare the intruder. The second can be outside the premises which will warn neighbors that something unusual is going on. Hopefully, your neighbors will investigate if they hear the siren.

Experts say that the siren is not enough. They suggest that a dual output system be arranged so that a siren sounds and house lights come on. This system is the best deterrent.

Using relays, you can operate the two output devices. Getting the lights on at the same time the siren sounds means that you have to make some branch circuit arrangements—such as paralleling some normal light switches and connecting cable extensions to a second relay to turn on *some lights*. You will probably have to be somewhat selective and just get a few lights on. It doesn't seem practical that you could get them all on at the same time without a major house rewiring job.

## SONIC LIGHT SWITCHES

There are some sonic light switches available. One unit plugs into a wall socket. When you depress a small "whistle" device, a supersonic (you can't hear it) tone is generated which then turns the light *on*. Repeating the procedure will turn the light *off*. Imagine a possibility with a siren and some of these units. If the siren can emit a supersonic tone which could trigger this type of light switch, then you could secure your home as the experts advise. If an intruder causes the siren to sound by entering or moving in a radiation field-protected room, then that sound would turn on lights in all areas within the range of the siren sound. No wires would then be needed to parallel light switches, and no relays would be required. These remote sonic switches only need be turned to some frequency which the siren can emit.

To find the frequency at which these "whistle" switches operate, you need a good high frequency amplifier, microphone and a good oscilloscope. You then cause the "whistle" part of the switch to emit the sound. Pick it up on the amplifier, feed it to the scope

Fig. 10-11. A mechanical siren is electrically operated.

383

and then turn the scope horizontal sweep control until you get a perfect (or nearly so) circle on the tube face. The frequency you then read on the horizontal frequency generating dial will be the same as the incoming frequency from the "whistle." Now you know what frequency the siren has to produce.

We like the supersonic control frequency because the receiver unit will not be triggered, normally, by audible sounds. You *could* use a lower frequency to turn on lights, using Radio Shack's voice-operated relay system. Certainly, it will be operated by a siren sound if it is loud enough. But that unit might also be operated by any other sound, and then would be likely be an irritation rather than a help. But we mention it in case you are electronically inclined and want to experiment.

## THE DOORBELL AND LIGHT

Perhaps you want to wire up a delay relay into the doorbell circuit so that this relay can turn on a light when the doorbell button is pushed. You would want the light to remain on even though someone depresses the doorbell button again and again, and you want the doorbell to ring each time the button switch is depressed. We suggest the circuit of Fig. 10-12.

You should have a long time delay on the relay so once it is energized, it will stay energized for at least one minute before it drops out of its own accord. Once it has been energized, the relay will just stay activated. But the bell or chime will not have this delay. Each time the circuit is completed, that chime or bell should sound. It's a handy idea and you may find it useful. The switch at A is to turn off the light in case you just want the doorbell (chime) to operate as it normally does.

## VIBRATION SENSORS

Vibration sensors can be "stuck" to the center of a window or door glass to cause an alarm by closing an internal switch, or opening an internal switch, if there is any kind of vibration to that glass. These units are wired in series with the rest of the circuit in the same manner that the magnetic switch units are wired. They might have some application to your security system planning, so be aware that they exist.

Another type of commercial vibration sensor, can sense footsteps, when units are placed in the ground at various intervals. We have examined some units which are used in highly classified

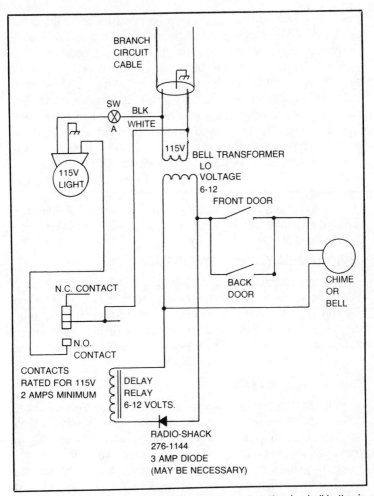

Fig. 10-12. Lights come on and the doorbell rings when the doorbell button is pushed.

security operations. They are small, compact and work exceptionally well.

## MECHANICAL VERSUS ELECTRONIC SWITCHES

You may be asking yourself why we suggested the magnetic type switch on windows and doors when there are many lever-action, mechanical switches which might be mounted to open or close—as desired—if a door is opened or a window is forced. We agree that you might use this type. But the actual mounting of the

385

mechanical switch will require a lot more precision to establish the necessary physical contact with the arm than is required in positioning the magnetic type switches. Also, if the mechical switch is activated frequently by the physical movement of the door or window, as it will be whether it is off or on, it may move slightly and its initial alignment may be thrown off. The magnetic switches must be close, but there is quite a bit of tolerance possible when mounting them. There is no physical contact with any part of the door or window which causes any movement of the mounting, as is the case in the lever switch type installation. The type units just seem to be better for security applications than the mechanical type switches.

If you happen to have some good micro-switches and want to install them, they will work out all right. Maintain the system by careful checking.

## PHOTOELECTRIC WARNING SYSTEMS

We will consider two photoelectric systems. One device turns on a light when the sun goes down and turns that light off when daylight occurs. The photoelectric beam system will sound an alarm whenever anyone breaks the beam between its light source and its receiver photocell. Both types are readily available on the market from places like Radio Shack.

### Photocell Lighting System

The photocell-operated lighting system is a good one for many applications. Some units simply screw into a lamp socket; you then screw the light bulb into them. This insures that a light comes on in your residence even though you are not there, and thus prevents you from going into a dark residence. Such a unit can also be installed in the light socket over the outside doorway so that it comes on and floods the entry with light.

The disadvantage of this kind of unit is that, because of its regularity, an intruder might determine that it is a light-controlled switch device. He might then unscrew the light bulb on the outside at least and put you in the dark when arriving home. You might put a metal grill protective cover over the unit to prevent removal of the bulb. We recommend that you change bulbs regularly and check the unit often to see that it is working all right.

### Photoelectric Beam System

The photoelectric beam type of alarm system is not widely popular for night protection because its beam might easily be seen in the darkness. This type relies on having that beam focused into the electronic eye of the receiving unit so that it hits the photo unit inside and causes activation of a relay (Fig. 10-13B). The relay, in turn, is wired to some light, siren, bell or other alarm, which sounds if the light beam is broken.

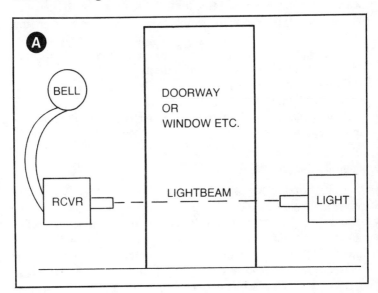

Fig. 10-13. Photoelectric warning systems. (A) Installation of the photoelectric eye. (B) Actual beam type photoelectric devices.

These units, diagrammed in Fig. 10-13A, are often used at below knee level so they won't be noticed so readily. They work well in daylight as well as night hours because of the cylinder hooding of the photocell on the receiver and the focused beam from the light source. The units use 115 volts as a primary energy source, and you just plug into outlets for their use. The relay can operate any type alarm by simply wiring to terminals on the receiver.

Some security personnel feel these units are a good addition to other systems because you cannot disarm them easily unless you are quite knowledgeable in electronics. Of course, one might use a strong flashlight focused onto the receiver cylinder while he slips past the regular beam. The flashlight beam might hold the receiver unit relay from activating.

### Keep the Sensors Hidden

The crux of the situation in security systems is that the sensors should be hidden so they are not easy to detect. They should not be easily disarmed. Many systems might be used. One system, for example, might be set as an operating decoy. When an intruder disables it, he thinks he is safe and then runs into the second system without knowing it is there. Some systems might have loud, audible sounds coming from them. We discuss this possibility in connection with a large *sentry robot* in The Complete Handbook of Robotics (TAB book No. 1071). We also discuss the use of robots as security guards in that book.

It is mainly up to you to figure out how to use these alarm systems for best protection. We highly recommend the photoelectric operated light for anyone who is living alone. Several might be useful. If you live in such an environment where you cannot get the required amount of daylight to turn the thing off during the day, then we suggest several "timed" light control devices which turn on lights at a given time, and turn them off at a second time. Both times are set by you.

### WIRING FOR A REMOTE CONTROLLED LIGHT

Figure 10-14 shows how to connect two lights in parallel, like a security light which is some distance from a hall light. Both lights will go on and off simultaneously.

Switch 1 is the wall switch normally used to control an overhead hall switch or room light. Run a new section of Romex to the

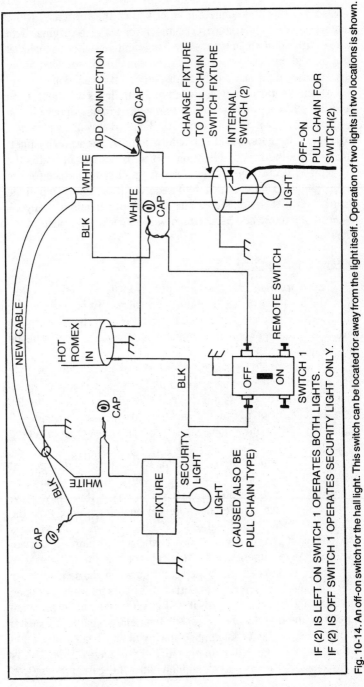

Fig. 10-14. An off-on switch for the hall light. This switch can be located for away from the light itself. Operation of two lights in two locations is shown.

IF (2) IS LEFT ON SWITCH 1 OPERATES BOTH LIGHTS.
IF (2) IS OFF SWITCH 1 OPERATES SECURITY LIGHT ONLY.

389

light socket box and connect it right across the terminals which connect to the light fixture. You must replace the original light fixture with a pull-chain type which has a built-in switch to turn that light off and on. The new Romex then runs to the location of the new security light and connects directly to it as shown.

When you turn the wall switch on, both lights will light if you have the pull chain switch on. If this switch is off, then only the security light switch will come on. Thus, the wall switch now controls both lights, but it especially controls the security light. You can turn the security light off so the hall switch operates the hall light as it normally did by getting a pull-chain fixture for the security light. That will turn it off and keep it off until you pull the chain to set it to the on state. The danger with this is that anyone else can also pull the chain, which will defeat the purpose of the security light.

## GARAGE DOOR OPENER

There are many garage door opening systems on the market. They all operate with a little remote control generating transmitter which sends signals to a permanently on receiver. The receiver is a part of the receiver control mechanism unit which is located inside the garage.

Some of these systems have a relatively simple code for causing the door to open and shut. But others have a more complex code which can be changed if desired. We like the more complex code as it will not permit the system to respond to accidental signals, or even to intentional signals by someone trying to get the door open.

A good, medium priced system, with a three digit code (at least, is the least likely to respond to spurious signals and to be operated by someone who doesn't know the code. If you have, say, seven dials which can be set to get an operating code, then the number of possible combinations and permutations is $2^7$, which is 128 for the off-on type digital code. That is a large number to try to go through experimentally. So this type of system is the safest.

There should be lights which come on when you send the door opening signal. A light on the control unit lights up when you open the door, and in some systems it will go off automatically a certain time after the door has been closed. We believe that there should also be a second light (floodlight) which will illuminate the outside of the garage and keep illuminating it after you drive inside. This is to *discourage* someone from slipping into the garage when you drive inside, as one might be able to do when the car lights go inside and leave the outside dark. A simple parallel circuit arrangement will do the job quite easily. Actually you can extend the lighting so up to three lights are turned on.

The garage door systems are quite easy to install. You must have access to a 115 volt ac receptacle outlet or to a light box in the garage which can furnish this power. Remember, if you must connect to an existing light, it probably will be turned off and on with a switch. You must turn the switch on and leave it that way all the time. A pull-chain arrangement for the garage light, then, might be used to give you light when working there.

If you can run a new line of 12/2 from below the switch box (or connect to the white and black lines in the switch box) to a new box near the door opener, this is best. Then no one can turn the device off, either accidentally or on purpose, at any time. Also, you will use the regular light switch to control the normal garage light.

## AUDIBLE AND VISUAL MONITORING SYSTEMS

There are two ways in which this monitoring can be accomplished. The first is quite easy. Simply buy two FM type *carrier current* intercom stations, such as found at Radio Shack. Place one in the room area to be monitored with its "send" switch depressed and its volume set very high so it is extremely sensitive. Simply plug it into an existing receptacle and the signals go over the branch circuit lines.

You then place the second unit—sometimes as far away as a neighbor's house—and put it on "receive." Turn its volume up high. You can hear every sound in the room being monitored and in several nearby rooms if the doors are left open. This receiver unit simply plugs into a wall receptacle, so there are no wires to contend with.

There may be, in the line, some kind of radio frequency blocking device which will prevent the reception of the carrier current radio frequency being generated in the transmitting unit. You just have to check that out. Second, you have to be careful to polarize the wall plugs on both units the same way, or you will get a buzz and noise on the receiver speaker. Plug the first unit into the wall. Then plug the second unit in and listen. If there is a buzz when the volume is turned up high, simply reverse the wall plug. That usually clears up the noise. You might have to try reversing the receiver unit. If that doesn't clear it up, try reversing the transmitter unit. Again, go back to the receiver unit and reverse its plug till you get the correct combination for clear reception.

If you get enough sound, you might use a voice-operated switch (kit available at Radio Shack) and let that turn on a light if there is any sound above a certain level. Don't let the battery run down. Most units operate on dry cell batteries. You might build a power supply or buy one which can operate the voice-operated relay unit and thus eliminate the need for battery worry.

The second monitoring system is, of course, television in a closed circuit arrangement. We are now talking about something which costs money. The installation requires coaxial cables and well-shielded power lines, plus the transmitter camera and the cable-connected receiver(s). Some of these units are programmed to turn at some given rate automatically. If an observer sees anything suspicious, he stops the scan, goes to manual control and zooms in or out on the object requiring attention. Other units require manual scan control. These are types which might be left in a given position of surveillance for a long time, and then moved by command as desired.

## BOOBY TRAP MAT SWITCH

We mentioned earlier that *Allied Electronics* has available a mat switch which can be used in security system planning and installation. These are operated with a 24 volt base and can operate a relay which will control any 115 volt device such as a light or siren. You might install one or several of these mats on the floor at vulnerable points. You might even use them under windows in a home.

## THE BARKING DOG

We have shown how you can wire up a light to go on when the doorbell is rung. That light was operated by a relay. It is a simple operation to cause that relay to close the primary circuit to a home recorder—of the continuous rotating tape type—and cause an audible alarm to be emitted which the person at the door can understand. One such alarm is the *barking dog* sound. If you get a continuous recording of a barking dog, connect the playback recorder to the doorbell relay switch. The person ringing the doorbell will hear that dog barking and most likely will continue on his way if nobody answers the door.

## RADAR TYPE SENSORS

The development of the *radar* type sensor which is immune to normal audible noises, including automobiles and traffic, might be well adapted to your needs for an outside sensor. This unit can be mounted, inconspicuously, on a wall of the house, over a doorway or wherever. It will "watch" the area it faces for movement of any kind from anything within its range of sensor capability.

Since the units operate on radar frequencies, there is no sound from them or any indication that they have detected any motion within their radio frequency field or beam. But they will operate a solid state type relay device or a conventional mechanical relay when they make this detection. This can sound an audible alarm, turn on lights, or activate a silent alarm to a police station or neighbor's house. The range of this type unit may be controlled by adjustments.

## PLUG-IN WIRELESS REMOTE CONTROL SYSTEMS

On the market at present from Radio Shack and other stores are the remote operated switches in systems as shown in Fig. 10-15. This system has a master station, shown at center, with a push button coding face. By sending the proper code, you can control the activation of modules, shown at the right, which can control lights or other electrically operated devices from the master station's location. On the left of the master unit shown in Fig. 10-15 is a wall switch replacement unit, which then makes that wall switch a remotely controlled device. The unit shown is a dimmer switch, you can set the brightness of the light it controls.

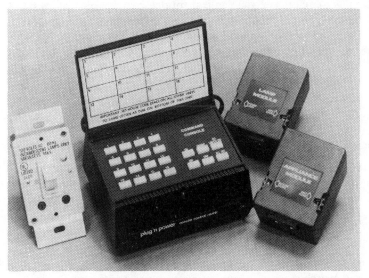

Fig. 10-15. Master and remote models for wireless control of electrical devices.

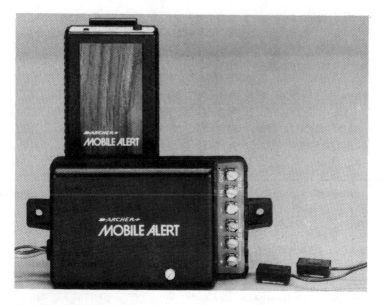

Fig. 10-16. A mobile alert system.

The master unit has 16 house code settings so that you can control many lights or devices remotely without any interaction between them. It is claimed that you can control up to 256 individual circuits, or you can control one circuit which has a lot of devices on it. The master unit has a code which you can send to cause all the lights in the house connected to remote modules to come on or go off as desired. With a system of this type, you have control remotely of lights or whatever. If you believe that a burglar is trying to enter your home, you can cause the lights to come on in whatever sequence you desire, to turn him away. But you must be there to activate the system; it does not operate automatically and by itself.

Regarding installation, simply plug a module into a wall receptacle and then plug your light, turned on, into the module. You plug your master unit into the wall receptacle near your bed, or wherever, and it is ready to use. If you replace the wall switches with the built-in type as shown on the left in Fig. 10-15, then disconnect the wires from the normal switch and replace it with one of these units. Be sure to turn off all the electricity to the branch circuit at the circuit breaker panel before you start disconnecting and connecting the switches.

## SECURITY RULES

We hope in this chapter we have given you some ideas as to how your home might be made more secure by using remotely controlled lights, automatic alarm systems, etc. You will probably envision other methods and systems which will help accomplish that same objective. If you are in doubt or need help, try consulting with security experts.

Remember that one of the best rules of security is not to let anyone know what you have done to protect your home, with the possible exception of showing the metallic ribbons on the windows

Fig. 10-17. A simple reliable off-on timer.

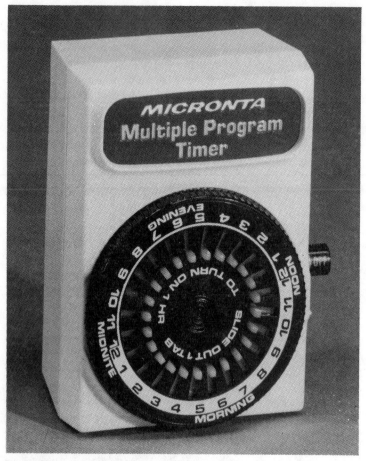

Fig. 10-18. A multiple program timer.

and placing small stickers on some windows stating that your home is protected. Some general rules of security, not necessarily involving any electrical devices, follow:

■ Keep a loud noisemaker at home. Use it if you need to attract attention or scare away a person.

■ Alert your neighbors to watch your home during the daylight, especially to see if anyone comes to your home and stays there.

■ *Never* let a stranger into the house, if they show some identification, check it out by calling their working place.

■ Don't tell people, other than trusted neighbors, that you are alone or leaving your house for a lengthy period.

■ Arrange to have a radio playing, lights on or other signs that someone is at home, even if you really aren't. Use a timer on these devices and lights.

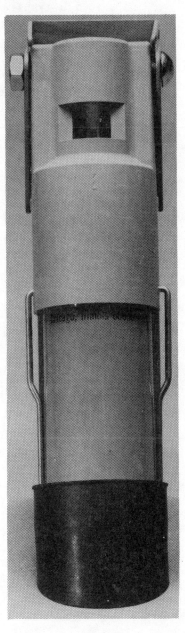

Fig. 10-19. This door alarm makes a loud noise.

- Arrange with the telephone company to route your calls to a friend's number while you are away on vacation.
- Never depend on just one light for protection. It might burn out. Always use two or more lights and have them on or activated by switches.

Figure 10-16 shows a mobile alert system. Use simple, reliable timers like those in Fig. 10-17 to turn various lights on and off while you are away. A more elaborate timer is shown in Fig. 10-18. Finally, if at home, you might use the door alarm shown in Fig. 10-19. It fastens to the door and blares out a warning when the door is pushed open.

# Wiring of Home Electronic Devices 11

When we consider the number and types of electrical and electronic devices which we find in a home nowadays, it seems like a good idea to examine in more detail how a home should be wired to accommodate these devices. We will look at some of the basic problems which can occur.

Once you have wiring in place, it's a good idea to test it before applying the full operating voltage and current to the branch circuit lines, or whatever lines are to contain the electricity. We have discussed this idea with some knowledgeable electricians. They to a large extent, may just rely on having done the wiring correctly the first time. They apply power and get their first indication that something is wrong when a circuit breaker trips, a fuse blows, or power doesn't appear somewhere that it should. Once they close the master circuit breaker, with all other circuit breakers set to the on position, they are alert to see if any breaker trips. If it does not, then they go through the house and turn on each circuit, one at a time, testing it for its ability to light a lamp, operate a fan, start the air conditioner or whatever. If everything works out all right and power is where it should be, the electricians then say the job is done and the electrical system is ready for use.

You could do that also when adding new circuits. Just put them in, connect them, and then turn on the circuit breaker which energizes them and see if you get power. We have to admit that this is the "quick and dirty" way to do it. Maybe it is the best way when you have lots of wiring to check such as when wiring a new complex of homes.

## WIRING MISTAKES

Also, you must consider the answer to the question, "How many mistakes can I possibly make?" There really won't be very many. If you run Romex cable, you might cross wire the black and white wires somewhere, causing a short circuit. That would cause the tripping of a breaker, perhaps resulting in the burning out of the breaker. You might not have a good connection somewhere and that would cause the loss of power in some circuit, which you will find as you test out each one individually. Some switches may be wired so that they do not exactly control the device they are supposed to. For example, if you have two switches in a given box, it really doesn't make any different which one lights the overhead light, or operates the garbage disposal. The switches must be capable of handling the power for the device they control.

If you have runs through metallic conduit, you might have "grounds" caused by the scraping off of the wire insulation at some joint or box so that the wire itself touches the box's metallic part. This might not cause any circuit breaker tripping or any loss of power. It just might make that metallic conduit "hot," and dangerous. Since the "common" line of a home system is grounded, probably nothing would happen if any insulation was scraped from a common line so it touched the metallic part of a conduit or box. If the insulation was scraped from a black or "hot" wire, you'd have a short circuit and circuit breaker tripping right away.

## TESTING OF ELECTRICAL CIRCUITS

We assume that you have a home that has been wired with no problems arising from that wiring except, perhaps, that it is now overloaded or has insufficient outlets. You can make an expansion as we have shown and explained previously. Now you want to test it. Let us assume the most basic possible situation where you have simply run a Romex cable from the circuit breaker panel to the location of some new receptacles. You have left each end of the cable free and not connected to anything. Now you want to test your cable run or wires in a conduit for continuity, short circuits and grounding. You can make up a little test lamp device by using a few batteries and a light as shown in Figs. 11-1A through 11-1C.

Connect the black and white wires together at the box or switch end of the Romex as shown at B. Apply the probes to 1 and 2

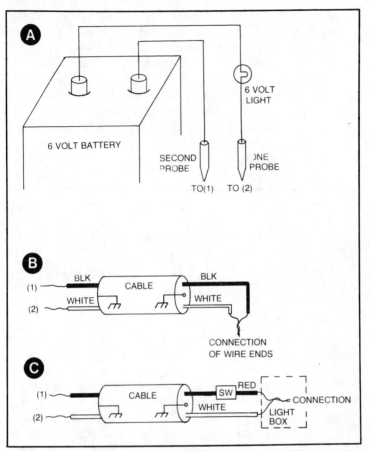

Fig. 11-1. Some wiring testing circuits. (A) The simple test unit. (B) A continuity test. (C) Testing the switch, cable and device.

at the circuit breaker panel end of the Romex (or wires in a conduit). You should have your indicator light burn brightly as it normally does when you touch the tips of the probe wires together to test your lamp-battery system.

If your "run" has a switch in it as shown at C, then you connect the two ends of the wires together past the switch and turn the switch on. Again, you should have your light burn brightly when the probes are connected to 1 and 2. Your test light should go off if you place the switch in the off position. That tests the switch capability, also.

This type of testing is called the series test. You need to make one more. Connect one of the probes to a wire (white or black) and the other one to the ground wire. There should not be any burning of your light. Now connect from the second wire to that same ground wire. Again, there should be no light. Now connect one probe to a wire and the second probe to the metallic part of a conduit, if used, or to any nearby water or gas pipe. There should be no light burning from your test lamp. You have now tested for a short circuit of the direct short type. It is possible that you could have a high-resistance short, but you will not be able to test for this without a good sensitive volt-ohm meter used in the same manner as your test light. Normally, you do not have to test for a high resistance short unless you have had undue difficulty running the wires in conduit or other places where you have a "feeling" the wires might have had some insulation removed.

Figure 11-2 shows one case where the cable might have run past a couple of pipes. When pulling it through behind a sheetrock wall or gypsum wall, it's not hard to scrape insulation from the wires as the wires are pulled over the pipes. Pipes are grounded of course.

## OPEN CIRCUITS

Suppose that you make the continuity test and find that your line is not continuous. What then? The first thing you do is to check your own connections. Be sure you have scraped the insulation away from the wires you have connected together and that those wires are firmly joined in a good electrical connection. Sometimes, in a hurry, people do not scrape the wires until they are really bare. They leave some deposit on the wires which can actually insulate them against such voltages as are being used for tests. Check the probe connections, the battery connections, the light connections, the wire connections at the end of the run, and the switch connections if one is in the circuit. Be sure none of these is at fault.

If you find no fault with your connections, then you may assume that something went wrong in the "run." That could mean doing the job over. You can pull out sections of the line (cable) or wires and examine them to see where and if they are broken or to find out what happened. No continuity may indicate a break somewhere. Then you have to run new wires or cable.

Another example of how you might have trouble with a nonmetallic sheathed cable (Romex) is shown in Fig. 11-3. Here the cable has been pulled past a metal gusset in a roof section. If you don't think that the sharp edge of that gusset can scrape away the cable or even break the wire inside, you are wrong. It certainly may

Fig. 11-2. If insulation is scraped away and the conductor touches metal pipe, a ground will result.

do that. Then you are faced with a problem of running down that "open" in the cable and trying to get a new one in place.

There is also the danger that the gusset or some sharp edge will scrape away the insulation and some of the wire's metal. The wire size at that point is reduced from a size 12 to a size 20. Now you test the circuit and find it seems to be okay. However, when that circuit is fully loaded, the wire will heat and separate and you'll have another headache to contend with. The running of nonmetallic sheathed cable (Romex) inside walls to expand wiring or to rewire homes should be done very carefully to insure that you won't have to take down a wall to repair some damage caused by using force to pull the cables through. Use common sense and save yourself money and time.

### Replacing a Bad Wire In Conduit

If the wires are run in conduit, you still might have that problem of having a wire scraped to reduce its size. We actually experienced that situation and didn't know there was a problem until we fully loaded a circuit one time and had smoke come out of the circuit breaker box conduit pipe. The wire separated, the circuit breaker tripped, and it was a real chore to pull out the bad wire. Fortunately there was only one. Using "grease" a tight "fishline" connection, we carefully worked the fishline through the conduit and pulled new wire into the conduit.

This incident illustrates how you handle the replacement of a bad wire in conduit. The use of "grease" gives the wire good slippage and helps it get through the mess of other wires which might be present in the conduit. Pushing the fishline through the conduit to pull the wire through takes a little doing and care. You do not want that metal fishline to scrape away insulation, either. It takes a lot of patience and perseverance.

So we treat the open circuit which should have been a closed and continuous circuit and we find out what is wrong. In the worst case, we run a new cable or wires to replace the bad section. Now we think about having a circuit which should test "open" and it tests closed, meaning it has a short circuit on it. Isn't it nice that we have so many challenging problems to tickle our fancy in electrical wiring? There's never a dull moment.

### Open and Continuity Tests

The branch circuits to receptacles should not test "closed" or continuous unless you have put a shorting plug in a receptacle on

Fig. 11-3. Pulling cable past a metal gusset may cause trouble.

405

that branch line. The receptacle circuit should show an "open" when tested from the circuit breaker panel end of the cable. This also applies to any kind of outlet which is of a type where something plugs into it. The "open" circuit test means that our test light will not light when connected to the circuit breaker ends of the cable. Nor will it light when we test from each line of the cable to ground, or a metallic conduit if wires and conduit are used instead of the nonmetallic sheathed cable (Romex).

But the lines should not touch anything. There should be no way to get that lamp lit except by shorting the cable ends at one end or the other and testing from the opposite end. You want an "open" circuit test in this case. But it may not be. You might have a break in the line or bad receptacle connections which would also cause the "*open*" to show up on the test.

If you follow the "open test" with a continuity test, shorting one end of the line as we have described earlier, you may discover a problem. Correct it by being systematic. First, check all receptacles to be sure they are okay. Then plug a light or lamp into them. Sometimes you get a contrary receptacle which has its metal fingers so bent that it won't make contact. It needs to be replaced. Next, check all screw terminal connections. You may find that a wire has slipped from a terminal screw or that someone screwed that terminal too tightly and broke the wire right at the screw. Correct these situations and then suspect troubles with the cable or wiring itself.

One sure test that many electricians make is to get a temporary permit to have the electric company connect up power to the house. Then they turn everything on. If there is any circuit which isn't working, they find out about it right away. They test each outlet for power and each light and switch for correct operation. There is no further testing unless a branch circuit doesn't give the power it should. Of course, this check doesn't test branch circuits for full loads and possible overloads which might occur. But it does provide information as to whether the place has electricity or not.

Sometimes switches or fixtures will not work properly and then you must replace them. In general, if the work has been done properly, neatly, professionally and carefully, you find that there are few problems to solve after the job is finished. Look at the messy, unprofessional job in Fig. 11-4.

## EXTENDING THE HI-FI SPEAKER SYSTEM

If you are building a home, then you may have already planned for an additional cable, of the telephone type, which can connect to

Fig. 11-4. This wiring job is very poor.

speakers in each bedroom or other designated rooms. You might want to have the output of a central hi-fi system "piped" to these rooms for increased living pleasure. You may want to include the speakers themselves, as ceiling units, as a part of this installation. Or you might want an outlet terminal to connect to some special type and kind of speaker at a particular location.

There may be some precautions which you should follow in providing this type of installation. First, you will not want your speaker cable near any other electrical cables. Second, if you think there might be a possibility that you would want to convert the speaker terminal into an intercom type terminal at any time, you should have well-shielded cable, perhaps three-wire type. Probably the most secure (from electrical pickup interference) cable would be a coaxial type, but it is relatively expensive.

Experts in sound systems say if a run to a speaker is over 20 to 25 feet, then you should use impedance-matching transformers at each end of the run or cable length. The use of these small transformers can prevent loss of frequencies, meaning higher fidelity. They can give a better power transfer from the master station so that you do not overload your amplifiers to get adequate sound at the ends of the cables. You must be careful, though, not to locate any impedance-matching transformer where it will be subject to an electric or magnetic field from some other type of transformer or motor. It can pick up the "hum" and give you distortion in this case.

Usually these transformers are located right on the speakers themselves which prevents trouble.

If the sound system is a multi-speaker type, then the impedance-matching transformers will be somewhat different in value than those for a single extension speaker. It is well and good to consult with your local radio shop about matching impedances on speakers for sound or intercom systems to be sure you are on the right track. One thing you should not do is to try and run extension wires from the amplifier to remotely located speakers, if they re over 20 to 25 feet away, without using the matching transformer system.

When you use the impedance-matching transformers, you can utilize normal cables of the speaker extension type without too many problems. Just keep the line away from other electrical cables. The fact that you will be converting the line impedance to a higher value and actually feeding more power over it when it is properly matched will help to reduce hum, distortion and other problems. You will find that it takes less power from your amplifier to get good volume at the far end when you have the system properly impedance-matched. This may mean that your amplifier will need a new output transformer or an auxiliary one if it is not designed to feed into a multi-impedance output. Figures 11-5 A and 11-5B shows the general concept of wiring for this type of sound or intercom system layout.

We show the use of 250 ohm impedance-matching transformers in Fig. 11-5. In actual practice, it is necessary to consider the number of transformers and speakers on a line. Then calculate the impedance to be matched by the amplifier transformer. It will be less than the 250 ohms shown, possible 125 ohms, For one or two extension speakers, you might get a pretty good match as shown. The best way to determine whether you have good matching is to consider how high you have to raise the volume of the amplifier to get a good loud sound from each extension speaker. If that level must be very high, then you might have a bad impedance-matching system. Again, we advise consulting with a good radio or sound shop for particulars before you try to install a remote speaker or intercom system.

## INSTALLATION OF TELEVISION OUTLETS IN VARIOUS ROOMS

Today many people want their own telephone and television in their room. That can mean a lot of television antennas outside, or it can mean one good television antenna system and feeds to a lot of outlets inside the home. A good television supply store will tell you about a multiple matching system for feeding the output of one antenna to lots of rooms. They will advise you on the type of coaxial cable used to carry this signal, clearly and without interference, from the matching device to the rooms. Follow their instructions to

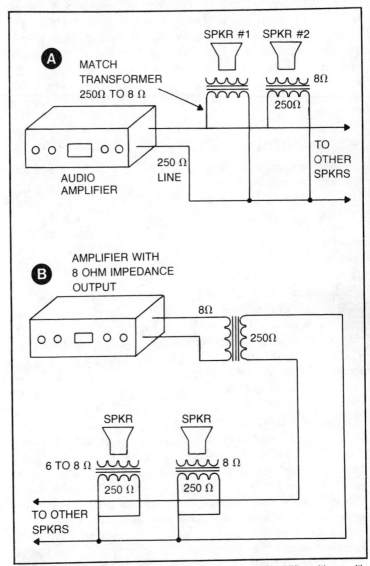

Fig. 11-5. Using impedance-matching transformers. (A) Amplifier with a multi-impedance output . (B) Amplifier with an 8 ohm impedance.

the letter. Be sure to keep your television outlet away from electrical outlets as a matter of general precaution. You won't pick up hum at the signal frequency. In some cases you might even get a better signal by having some pickup from the electric branch circuit lines, but it isn't the best way to do it.

Having each electrical system separate, complete and distinct from the others is the best way to insure a trouble free and professional type installation. If you will look back at Fig. 6-52, you will see an example of many coaxial leads from a television antenna system which will be connected to a matching transformer-distribution box. This, in turn, is connected to an antenna. The other ends of these cables terminate in outlets in various rooms of the building.

## INTERCOM SYSTEM

Modern homes use intercoms which can provide communication with anyone at a front or back door of the home. These two stations are sometimes just part of a more elaborate system which connects to the garage, workshop, attic and bedrooms, if all that is desired. It is wise to consider the use of such a system and then plan wiring for it, even if it is not used at all or just partially used. At least, if the system is installed, it is there in case you later want to use it.

Intercoms use a relatively light type of cable, just about the same as used in a telephone extension line. There may have to be three or more wires in the cable. That cable may have to be shielded well because you use the main amplifier, located somewhere in the house, to amplify the sound coming from the outside speaker. So the lead from the outside speaker to the amplifier is very sensitive to any kind of disturbance pickup such as hum, electrical sparking noises and air conditioner belt static.

We have also found that the grounding of the shield on the cable can be a very touchy task. If the cable is grounded to water pipes at each end and the ends are a long way apart, sometimes a ground voltage loop will develop which puts a small voltage on the shield between the two points of grounding. This can be especially true if the water pipe system comes in contact with earth and thus gets some electrolysis, or if it has several segments of different metals in its structure. This grounding system voltage may also be induced from several connections to electrical cable ground wires or where common wires (white ones) may be grounded through some, perhaps, unplanned situation.

It is therefore best to try to have one grounding point for the cable preferably a good water pipe if it goes into the earth in a metallic form. If the cable goes into a plastic pipe (PVC) as it leaves the house, then it is not a good ground. A grounding rod system must be used. You should plan to have the cable installed for this type of intercom installation, especially if you are building a home or residence. Apartments have such systems, usually, and condominiums do also.

## AC-DC TRANSFORMER TYPE EQUIPMENT

It becomes more and more important to observe the polarity of the receptacle plug with the advent of more types of electronic devices which operate on ac-dc. This is not so important if the equipment has a transformer to furnish the necessary equipment power. Yet, it can be important even then for such devices as televisions or tape recorders which use a good grounding system to prevent transients from causing damage. Computing equipment for home operation may also be very sensitive to polarity of plug operation.

In a well designed and professionally wired home, you will find that a receptacle has one large size plug insert hole and one smaller size hole, in addition to the circular ground connection hole. See Fig. 11-6. The electronic equipment which requires observance of the polarity of this receptacle will have a larger size "finger" and a smaller size" finger. Thus, you can't get the plug into the receptacle in the wrong direction.

When the receptacles are wired properly, you will find that the white, common wire is connected to the "silver" side of the receptacle. The black wire is then connected to the gold side of the receptacle. This will give the proper polarity to the receptacle socket as shown in Fig. 11-6. Be sure the receptacles are not wired so that the polarity is reversed. You can do nothing about that except to inspect the receptacle connections by removing the wall plate and the screws holding the receptacle in place and seeing if the wiring is correct. If it is not, then turn off the power and reverse them. You might suspect this situation if you have electronic equipment which has a polarized plug, but doesn't seem to be working properly in your residence.

Some equipment with transformer power supplies use a polarized plug to get a proper grounding for the system. If the manufacturer finds it necessary to ground the equipment through polarization of the electrical plug, then it must be necessary for the correct operation of that equipment or the protection of that equipment from voltage transients. Never replace a polarized plug with just any old plug. If you have to replace one, be sure you observe the wiring. Make sure your connections to a new polarized plug are the same as for the old plug.

## THERMOSTAT CONTROLS

There are some types of thermostat controls which claim to have "brains". These are just automatic timing devices, which use an electric clock type mechanism to turn on and off the heating and/or cooling at various times inside the residence. This is good, of course, because you can set the heat higher for day than at night, or set the cooling cycle to start at midday instead of early morning.

### Wiring For the Timing Units

Have the wires to the furnace-cooler unit valves and motors coming down from the attic or up from the basement. You will need a line from the house branch current nearest the unit location to power the clockwork mechanism. In most thermostat control devices you do not need 115 volts ac, so there is no line run for that purpose. What you have to add is this nonmetallic sheathed (Romex) cable to get some "juice" at the thermostat location to power its operating equipment, normally a clock timer device of some kind.

If you have a switch nearby, you might look inside that box and see if you can tap into the black and white wires coming into that box. If you have a receptacle either on the same wall or in a room opposite the thermostat, you may be able to parallel the wire connections to the receptacle and run the cable through the wall and out at the new thermostat location. You may find that the shortest and best run may be from overhead fixture down to the thermostat location. This will be good only if there is power in the ceiling fixture socket which is not switch-controlled. If the power there is switch-controlled, then you cannot use it.

Before replacing any conventional thermostat with one of these new "brain" types, be sure to look each one over carefully. Find out if the new one has the required connections—all of them—to connect to all the wires of the old thermostat. If it does not have all the connections, then you'd better by careful as there will be some element of the heating-cooling system which will not operate by the new device. If a direct replacement can be made and the only additional wiring will be the 115 volt ac line, then replace

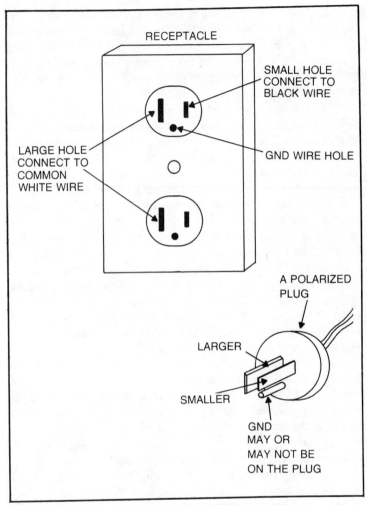

RECEPTACLE

SMALL HOLE
CONNECT TO
BLACK WIRE

LARGE HOLE
CONNECT TO
COMMON
WHITE WIRE

GND WIRE HOLE

A POLARIZED
PLUG

LARGER

SMALLER

GND
MAY OR
MAY NOT BE
ON THE PLUG

Fig. 11-6. This wall receptacle has polarized holes for the plug.

the old unit if you so desire. Perhaps you'll save something on your utility bills by doing so.

## Computerized Thermostat Controls

This system is more complicated. Computerized thermostat control adjusts can the heat in each room as you desire. But this system has to be put in when the original heating-cooling system is installed. You need a *sensor* which is a temperature gauge in each room that senses the temperature in that room. Second, you will need a means of mechanically and physically shutting off or letting the heat or cool air come into that room or location. A motorized device or flapper in the tunnels or ducts can be used to heating and cooling from a central location to each room or area. That means a lot of control wires coming to some central location where the "computer brain" is to be located.

The computer brain, in turn, must be able to turn on and off switches to control the flappers or dampers in the ducts. It must be able to make a comparison of the heat or cold as monitored by the temperature gauge in each room, with some standard value which you can control or set. Now we have the elements of the computerized control system which will be necessary.

The computer will sense a voltage level which is directly related to the room or area temperature. It will then compare this to a voltage level that you have set which, in turn, states the temperature to be maintained in that room or location. If it is too cold, the computer comparison unit, sensing that the temperature voltage is too low, will cause a motor to run which opens a damper or flapper. It will start the heating motor and system. Warm air will come into the room and area and cause the temperature to rise. The sensor voltage will rise until the computer knows it has the value you set. Then it turns off the heating unit.

Now that is just for one room. Imagine the operation if you have five rooms to monitor and control, or even ten rooms. Lots of wiring will be needed. That is why we suggest it should be done when the residence is built. If it is done later, the task will be harder and more expensive. Computerized thermostats will be found frequently in residences and offices in the near future because people are interested in saving energy.

The wires needed for control of the various elements of this system, and the wiring to the sensors, need not be electrical cable necessarily. You can use the much smaller "telephone cable" type of line, which could be a multiple cable with "dropoffs" wherever a sensor or a control motor are to be located. The lines to the control motors from the computer will be small in size, perhaps #18 gauge or so. But these will then control relays of either the heavy duty mechanical types or solid state types which, in turn, will control the application of the real power, 115 volts ac to the flapper motors or moving devices. Of course, you will have to have branch circuits to each flapper motor location. These lines will come from your central circuit breaker box or the fuse box.

You will need the control lines which handle small voltages

and run from the computer to the various areas. Also, you will need the branch circuit lines similar to those used for applying electrical current to the kitchen appliances, running from the circuit breaker source to the various motorized control units. Heavy duty relays are required which should be sealed. You will need temperature gauge type sensors and the control motors. These control motors will have limit switches on them so that they cannot keep running after they open a flapper damper. They will use these same types of limit switches to remove power when they close the damper. The running of the motor must be restricted in both directions of motion. See *Model Radio Control—3rd Edition* (TAB book No. 1174) for more information on limit switches and self neutralizing switches.

## PROBLEMS WITH LAMPS

Let's consider lamps first. They are the easiest devices to repair and maintain, and they are probably the most commonly used in the house. Lamps can have several types of problems. First, the bulb may be burned out. Second, it may not be plugged into the wall. Third, the cord at the lamp base or at the plug end may be broken, twisted into two parts, or twisted so that the insulation breaks away and the wires short together. Then the switch may be faulty and not operate as it should. The contacts in the lamp light socket may get bent down so they do not make connection to the bulb. Following are some suggestions for finding a solution to lamp problems.

**Step 1.** Try another bulb. You'd be surprised how many persons may take a lamp apart when all that is wrong is the bulb.

**Step 2.** Check the wall plug. It may *look like* it is firmly seated in the wall socket, but is may not be making a connection. Its prongs may be bent so they do not make contact. The wall socket may have enlarged holes or be worn out.

**Step 3.** Now examine the cord at the base of the lamp and at the plug end. If it looks cracked or broken, then replace it. You can buy a length of the same type of cord in almost any electrical department. If you examine the plug connections carefully, you can duplicate them with the new cable. Then, with some experimentation, you can find out how to take the lamp apart and run the new cable where the old one went.

**Step 4.** The plug fits, the bulb is okay and the lamp cord is all right. But the light won't work. Then you might suspect the switch.

This means taking the upper part of the lamp apart, removing the switch, replacing it with a duplicate, and putting the lamp back together. Do not try to do this when the lamp is plugged into a wall socket. Try to get an exact replacement switch.

**Step 5.** Before replacing the switch, you might check the socket contacts. Sometimes they get bent and do not make connection to the bulb base. Just disconnect the plug from the wall and use a small screwdriver or nailfile to pry up the contacts a little, so they are high enough to make good connection to the bulb.

## USING STEP-BY-STEP PROCEDURES

These steps are somewhat significant when you have troubles with any home appliances or electrical devices. A logical step-by-step procedure in locating troubles is necessary. We might have even suggested that when a lamp does not light, it might be well to plug another lamp into that very socket and see if it lights. That way you can check the house branch circuit without going outside, into the garage or wherever to check the circuit breaker on that branch circuit.

Many home electrical problems are really so simple. A plug which doesn't fit firmly into the wall socket or is not plugged into the socket all the way is common. Only as a last resort should you attempt to take the device apart to check its internal parts and wiring. Most of the time it takes an expert to do that and get it back together properly. If you take the device apart, you probably won't have the repair parts if they are needed. Make all the checks we have listed and then, if you are not expert in such matters, visit your local repair shop for help.

## CENTER-TRIP CIRCUIT BREAKERS

Many circuit breakers are simply an off-on type in which the reset lever moves from one extreme to the other on the case. Some *center trip* types have a little handle which goes to a central position when the device trips. That presents no difficulty, but what does present a problem occasionally is how to reset this type of circuit breaker. It is necessary to first move the reset lever to the full off position before moving it to the full on position, so that it will stay in the on position when you release it. If a circuit breaker in your home panel won't reset, then try moving it from full off to full on. It probably will then reset properly.

## PROBLEMS WITH FLUORESCENT LIGHTS

Fluorescent lights can be irritating to say the least. A fluorescent light may flicker on and off at various rates. It may hum, smoke and radiate considerable heat, or not light up as it should when you turn it on. Start isolating the problems by first checking the fuse. Take out the tube and clean the end contacts. When you put the tube back in its sockets, twist it several times so the tube pins actually wipe against the socket fingers and clean them.

Replace the ballast unit, which is the small fuse-like device that can be removed from the fixture by grasping firmly and twisting a half-turn or so. Sometimes the ballast unit is called the transformer, but we choose not to call the transformer by that name.

If the fixture is hot or humming, the transformer probably needs replacing. You have to take the fixture down to get at it as it is usually out of sight inside the structure. It may have some tar seeping out of it or other evidence that it has been running hot. Then you know it needs replacing. If it hums, but works, it also needs replacing. The tar has probably gone out of it and the transformer laminations are flapping around inside the case. Get an exact replacement. Connect it to the wires just as the old one was connected.

Try a new tube if the old one flickers excessively or doesn't light at all. Clean the old tubes with a damp rag while you have them out of their sockets. That will improve the light emission.

Check the fixture switch if all else seems okay.

## DOORBELL DIFFICULTIES

You push the doorbell button and it doesn't ring. What do you do next? See if the bell or chime itself is working properly. Sometimes lint, dust and dirt can accumulate in a chime or bell in such a way that the stroker or bar cannot move properly to sound the audible alert notes. You can free the stricker by cleaning the device with an electrical cleaning spray which you can get from radio parts houses, or a similar type of cleaning fluid or spray which will not harm the electrical wiring or circuits. You can easily tell if the chime striker is at fault when you try to move it by hand. In can be gummy and sticky and that tells you it needs cleaning. A doorbell or buzzer may have its contacts pitted and corroded so it doesn't vibrate as it should. You then replace the bell.

If the bell or chime is found to be okay, then check the doorbell switch button. Carefully remove it and look at the wires. Sometimes these are very short. If you are not careful, you might break them off inside the door jamb. So use care. Now you can test the switch button by simply shorting a wire, screwdriver blade or other metal across the switch connections. If the chime then sounds, and it doesn't when you press on the button, then you need to replace the switch button. It is a simple matter to unscrew the two connections to the old switch and connect them to a new one. Screw the new switch button back into its original location. Test it and you've done a good job when the chime or bell rings clearly.

Perhaps you could have a bad chime or bell. It does happen that the magnetic wiring of the device somehow fails and then it is defective. The only way to test that is to substitute another bell or chime and see if it works. Or get a voltmeter and test to see if, when you depress the doorbell button, that a voltage appears at the terminals of the chime or bell. It should be around 6 volts at least. If there is a partial short in the wiring or bell or chime, then that voltage may not be high enough to ring the device. If the device is bad, a voltage may be there and still it won't work. A substitute check may be necessary to find that out.

If you have the dead situation just mentioned, proceed to the attic or wherever the bell or chime transformer is located. Test it to see if it has burned out or become loose in its connection to the wiring or the hot branch circuit. Sometimes you can tell if it is operating by listening carefully to it. You might hear a slight hum as it is on all the time. If you do not hear this, then disconnect the bell wires and test the output terminals with a voltmeter. If you get no reading or a low reading less than 6 volts ac, replace the transformer. Be sure to turn the branch circuit breaker off before you attempt to disconnect the transformer primary from the house circuit "hot" lines.

We recall one situation where the doorbell didn't work. We traced the circuit until we located the transformer in a closet. Someone had apparently hit it breaking the connection to the light circuit hot line. We reconnected this broken connection and everything worked just fine. The problem here was trying to trace the circuit—the secondary circuit from the bell button back to the transformer. Sometimes that is not easy. Remember it is common practice to locate the bell transformer in the attic. If you fail to find one here, examine closets, pantries and other concealed places where it might be located. You might find it where you least expect it to be, attached closely to some light socket or outlet.

## AN AUTOMATIC LIGHT CIRCUIT

When poking around various dwellings, we came across a dandy little circuit used to turn on a closet light when the door is opened and to turn it off when the door is shut. The light requires a little circuit using a triac, a few resistors, and a reed switch operated by a magnet. The circuit is shown in Fig. 11-7.

The reed switch is located in the door jamb in such a way that when the magnet, located in the door edge, comes close to it the reed switch will open. When the magnet is pulled away, the reed switch will close. This is a normally closed type of magnetic reed switch.

Adjust the 200k potentiometer so that the triac will conduct when the door is open. Then leave it set that way. When you close the door, the reed switch opens and the light goes off. When you open the door, the light comes on. There is no physical pressure

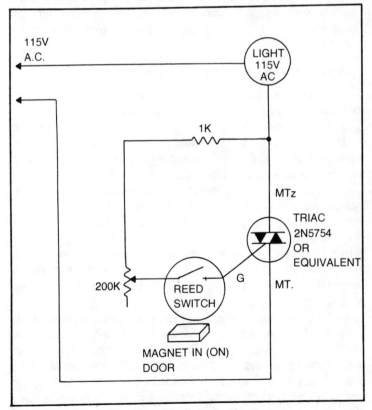

Fig. 11-7. A reed switch operated triac light control circuit.

required. If you place the magnet and reed switches carefully in the door edge and door jamb so they cannot be seen, you have a fine, concealed type of lighting system that will serve a very useful purpose. You will probably think of many other uses for this dandy little circuit.

## FLIP-FLOP OPERATED LIGHT-BUZZER SYSTEM

A very interesting circuit which causes both a light to burn and then a buzzer to make an audible sound is shown in Fig. 11-8. When B in Fig. 11-8 is depressed, the circuit is set to normal. If A in Fig. 11-8 is depressed, the left flip-flop is set and its high output will cause the light to burn after its relay or solid state control element are activated by the *or* gate output. If input A is activated again, a second time, then the flip-flop shifts and the output of the right one becomes high, causing both the light to burn and the buzzer to sound. If B is now activated, the whole circuit is reset back to the starting condition again. You can probably think of lots of different applications for this circuit, or one which is an expanded version of it.

## CONNECTING WIRES IN HOME BRANCH CIRCUITS

It used to be common to connect cable wires by scraping them carefully to remove all insulation and varnish from the wire ends. They were twisted together tightly and then soldered with a good resin core solder. That is not done any more. Now the method is to twist the wires and put on the insulating cap as we have described earlier. Let us expand on that idea further at this time.

If you have two or three wires to connect, measure the scraped ends and cut them so when the cap is screwed in place, no bare wire will show beneath it. This means about ½ inch of bare wire on each one will be used for the connection. Next twist the wires together clockwise so that when you screw the cap on them, the cap threads will tend to tighten the wires and will not attempt to loosen them. This works fine for two, three, or perhaps even four wires. Caps are available in all sizes.

If you have more wires to connect together in a common joint, then it is wise to get a *compression ring*. You twist the wires together as tightly as possible and then slip the compression ring over them. Crimp the ring down tightly on all wires, testing to be sure no single wire is still loose under the bond. Put on a cap, or carefully tape the connection with strong, electrical plastic tape. The best and preferred way is to use a plastic cap which can fit

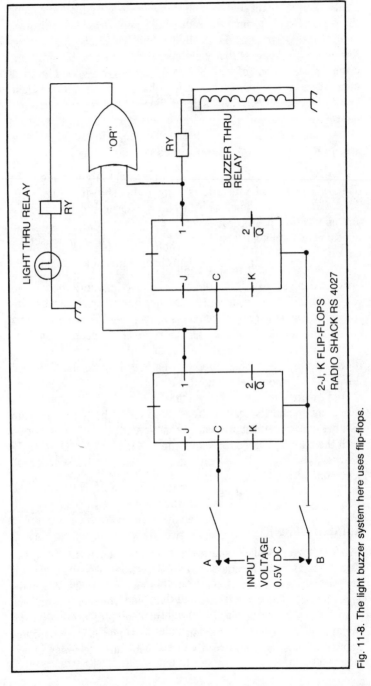

Fig. 11-8. The light buzzer system here uses flip-flops.

tightly over the joint so it won't become loose and fall off, exposing the wires. Bonding rings and caps of all sizes are available in store electrical departments. You will also find crimping tools in these locations. We suggest that you make some practice bindings and crimpings on scraps of wire before trying to get a good electrical connection on the real wiring. It isn't hard to do, but it does take some practice to get all wires tightly bonded and strongly fastened together.

## A CIRCUIT REQUIRING MANUAL RESET

A basic SCR (silicon controlled rectifier) circuit which will operate a relay when its gate is triggered is shown in Fig. 11-9. This circuit can be the basis for a burglar alarm. It can be used in any application where you want a sensor to cause a voltage to appear at an output and stay there, until you turn it off and reset the circuit.

The gate circuit in this application is biased so that it will prevent the SCR ( HEP R1103 ) from conducting, until the voltage applied to the gate is removed by opening one switch in series with the voltage and the gate. When this is done, the SCR will then start conducting and the gate loses all control over the conduction. The relay can then be energized. It, in turn, will cause a higher, more powerful voltage to be applied to whatever it controls. This end device will run or operate until the voltage to the SCR is interrupted by opening the reset switch. Once the SCR is turned off in this manner and the gate voltage is applied, the gate can again assume control. It can prevent conduction until a switch in series with the gate is opened, breaking the voltage line to the gate. By the way, you adjust the 50k potentiometer so that the SCR "fires" when one of the alarm switches is opened, and stops firing when the reset switch is opened.

## METALLIC CONDUIT

We have not discussed metallic conduit much in this work. Now let us examine it somewhat to find out how to use it. Consider first that it comes in various diameters and in lengths of usually about 10 feet. The conduit may be galvanized iron, but it usually is aluminum. When you use it, you want complete protection for wiring inside. When the conduit is sealed, you have protection against wetness and moisture which could cause troubles in the long run.

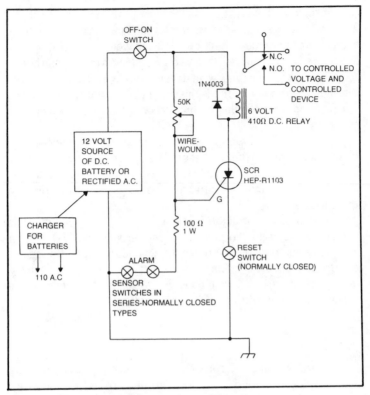

Fig. 11-9. This alarm circuit requires a manual reset.

## Pulling Through Electrical Wires

If you have to pull some electrical wires through that pipe, use a "fishline." It is a metallic ribbon which can bend easily around relatively large curves. In conduit, we should never find a bend which is smaller than eight times the diameter of the pipe. A small, strong, flexible fishline will be able to bend around that circuit or part of a circle as it is pushed into a conduit from end to end.

There is a definite number of wires which are permitted in a conduit under the National Electrical Code. See Table 11-1. The asterisks denote the wires most commonly used in home wiring.

A fishline can be made of steel, as we have previously indicated, or of strong plastic. The steel types may be from ⅛ to ¼ inch wide. The plastic types will have a width compatible with their tensile strength.

In any event there are various means of getting the wires through a section of rigid conduit be it metal or plastic. You may find, if the run is short, that you can just push the wires through. This may be accomplished in somewhat longer runs by getting some "conduit grease" or powder from your electrical supply house. You may even push "back" a single length of wire and attach to the three or four you want to pull through. Then pull them through this way. Sometimes, just a length of ordinary galvanized wire can be used for this pulling process instead of a length of copper wire.

When the run is long you should probably use the fishline. Push it back through the conduit until its end emerges. Fasten your wires to its hook section. Tape them securely after making a good mechanical connection. Gently, but firmly, pull the wires through the conduit.

If the run is a straight one, but over 10 feet in length, it is sometimes advisable to pull the wires through the first 10 foot section. Then route them through the second 10 foot section and so on. Don't fasten the coupling until you get all the wires through the pipe. There are good couplings for fastening the ends of the straight and bent section of conduit together.

## Using an Ell

When you come to a corner which you need to get around, an *ell* or corner connector does the job. Do not try to just pull the wires around the corner; you might break one or damage some by scraping off the insulation. The ell is a fitting with removable side or back which is screwed into place when on the conduit. When the back is removed, you have spaces for the incoming wires and for the placement of the ongoing wires. Change your fishline insertion point to a place further down the run and feed it back to the ell. There you make a new connection to the fishline hook, and carefully feed the wires into the new section of conduit as someone pulls firmly but gently on the fishline at the other end of the conduit run. When the wiring is all through, you fasten the side or back plate of the ell in position and all is tightly encompassed inside. Ells are used when going around a corner and you want to keep the conduit tight against the structure. Bends are used when you want to route the conduit from other than a straight line.

You must not have more than four 90 degree bends in one "run" of conduit. Get some help from an experienced electrician because bending is an art and takes quite a bit of skill. Of course, you can take your conduit and find out where your bends should be

Table 11-1. Number of Electrical Wires Permitted in Conduit.

| Wire Size | Conduit Size Diameter (inches) | | | | | |
|---|---|---|---|---|---|---|
| | ½ | ¾ | 1 | 1¼ | 1½ | 2 |
| 18 | 7 | 12 | 20 | 35 | 49 | 80 |
| 16 | 6 | 10 | 17 | 30 | 41 | 68 |
| 14 | 4 | 6 | 10 | 18 | 25 | 41 |
| *12 | 3 | 5 | 8 | 15 | 21 | 34 |
| *10 | 1 | 4 | 7 | 13 | 17 | 29 |
| *8 | - | 3 | 4 | 5 | | - |
| 6 | - | 2 | 3 | 4 | 5 | |

and what size they should be. Then have your electrician bend the conduit for you. You mount it and pull your wires through.

## Avoid Spliced Wires

The National Electrical Code says that you *must not have any spliced wires in a conduit*. That is a rule you should not voilate. If your wires are not long enough to go from one end of a conduit to the other end without splicing, and you don't want to buy more wire, consider using a box in the conduit where the splice may be permitted. Just cut your conduit, insert a box at the proper distance, run your wires to the box, splice them there, cap them, and continue the run with some more wire which will not be spliced inside the conduit.

## Mounting the Conduit

There isn't much problem using conduit for electrical wiring except, possibly, for the mounting of the conduit. Use proper straps which are available in electrical departments. Anchor the conduit firmly. Go around where people have used conduit and see how they have fastened it in place.

## Sheathed Cable

Remember, also, that there are various types of metallic sheathed cable, some of which have plastic insulation over the metallic, flexible sheathing. You may find you will want to use this cable instead of rigid conduit in many applications and locations. Examine fittings, boxes and connectors for this conduit and use them properly. Note, too, that the edges of the conduit can be sharp and will easily cut the insulation if you do not use a very careful method of feeding the wires into the conduit.

## RACEWAYS

Occasionally it is necessary to have a large tunnel known as a *raceway* to hold all the wiring used in a given location. Actually, the raceway is a rectangular box-like structure of metal. You won't find this, normally, in a home. However, you may find it in an apartment or conominium where there are concrete floors into which the wiring must run.

## MAKING RUNS IN FINISHED HOMES

When you add outlets, fixtures or switches in homes which are already constructed, you have a problem regarding how to get through walls, ceilings and down baseboards so that you can extend the wiring. There are many ways to approach the problem, depending on how the run is to be made.

Perhaps you are going to add a new receptacle some distance away, but in the same wall as an old receptacle. Then you might consider carefully removing the baseboard of the wall. Run the wire cable down from the old outlet until you can grasp it where the baseboard was. Feed it across the room, through holes or notches in the studs. Don't forget that if you do notch a stud, you must make it 1 ½ inches deep or cover it with a metal plate, so a nail from the outside won't puncture the cable. Get to the new location, go up the wall stud to the outlet opening and then replace the baseboard. Put in the outlets, make the connections and the job is done.

If you want to get to a ceiling fixture, you may have a more difficult problem. First, try to determine how the ceiling joists run—in which direction. You do this by making an opening at the fixture location. Then use a wire probe or a flexible rod of some kind to poke into the hole and locate the joists. Once you have located the direction they take, you can then feed a cable down such a joist to the wall. At the wall location, you then cut a small hole into the plaster so that you can "fish" around for the end of the cable. Pull it down. Now go into the wall itself about 1 foot below ceiling level and in line with the "run." Cut out another section of wallboard, say 6 inches square. You then want to drill a hole through the top support of the studs so you can feed the cable down through this hole into the wall. A long bit on the drill may be necessary.

Once you get the cable into the wall, pull it down to whatever level you need to reach. Again, you may have to make a cutout into the plasterboard near the bottom of the wall. Feed the cable down and then fish around for it until you can pull it out of the hole there.

Then you can pass it on to whatever location—say a receptacle location—that you might want to reach. If the cable goes to a switch location, then it won't go so far down and your cutout will be higher in the wall.

Remember that you will have to have a stud as anchor for your receptacle or switch box. You will have to have a cutout in the plasterboard at that location. It is easier to feed a cable down alongside a stud, and then find the end of the cable lower down, than it is to just let it go free into the space between studs. Try to make it follow a stud or joist whenever you push it into an inaccessible space.

We have investigated many situations where it became necessary to remove a whole section of wallboard to get an electrical wiring job accomplished. We hope yours won't be that bad. If it is, get professional help.

Hopefully, you now have a better understanding of what electrical wiring is all about. A do-it-yourselfer should be able to do many tasks related to electrical wiring; others are better left to professional electricians. At any rate, may your way always be bright and filled with the light of understanding. Keep your electrical connections secure, properly made and professional.

# Index

431